ENVIRONMENTAL REMEDIATION TECHNOLOGIES, REGULATIONS AND SAFETY

FLARING AND VENTING OF NATURAL GAS

BACKGROUND AND ISSUES, IN BRIEF

Environmental Remediation Technologies, Regulations and Safety

Additional books in this series can be found on Nova's website under the Series tab.

Additional e-books in this series can be found on Nova's website under the e-book tab.

ENVIRONMENTAL REMEDIATION TECHNOLOGIES, REGULATIONS AND SAFETY

FLARING AND VENTING OF NATURAL GAS

BACKGROUND AND ISSUES, IN BRIEF

LINDA R. HENDERSON
EDITOR

NOVINKA

New York

For permission to use material from this book please contact us:
Telephone 631-231-7269; Fax 631-231-8175
Web Site: http://www.novapublishers.com

Library of Congress Cataloging-in-Publication Data

ISBN: 978-1-63463-838-8

Published by Nova Science Publishers, Inc. † New York

CONTENTS

PREFACE

Natural gas is regularly lost as it is burned (flared) and released into the atmosphere (vented) during the production of oil and gas. While the exact amount of gas lost in this way is uncertain, given the vast extent of oil and gas production throughout the world, it could be significant. In addition to resulting in the loss of a potentially valuable resource, the burning and release of natural gas into the atmosphere contribute to greenhouse gas emissions, which are generally considered to be warming the earth's atmosphere. This book describes flaring and venting data and what the federal government could do to improve them; reports on the basis of available information, on the extent of flaring and venting and their contributions to greenhouse gases; and identifies opportunities for the federal government to reduce flaring and venting.

Chapter 1 – Natural gas is regularly lost as it is burned (flared) and released into the atmosphere (vented) during the production of oil and gas. GAO was asked to (1) describe flaring and venting data and what the federal government could do to improve them; (2) report, on the basis of available information, on the extent of flaring and venting and their contributions to greenhouse gases; and (3) identify opportunities for the federal government to reduce flaring and venting.

Chapter 2 – The Department of the Interior (Interior) leases public lands for oil and natural gas development, which generated about $9 billion in royalties in 2009. Some gas produced on these leases cannot be easily captured and is released (vented) directly to the atmosphere or is burned (flared). This vented and flared gas represents potential lost royalties for Interior and contributes to greenhouse gas emissions.

GAO was asked to (1) examine available estimates of the vented and flared natural gas on federal leases, (2) estimate the potential to capture additional gas with available technologies and associated potential increases in royalty payments and decreases in greenhouse gas emissions, and (3) assess the federal role in reducing venting and flaring. In addressing these objectives, GAO analyzed data from Interior, the Environmental Protection Agency (EPA), and others and interviewed agency and industry officials.

Chapter 3 – This is the Testimony of Andrew Logan, Director, Oil & Gas Industry Program, Ceres. Hearing on "Benefits of and Challenges to Energy Access in the 21st Century: Fuel Supply and Infrastructure.'"

In: Flaring and Venting of Natural Gas
Editor: Linda R. Henderson
ISBN: 978-1-63463-838-8

Chapter 1

NATURAL GAS FLARING AND VENTING: OPPORTUNITIES TO IMPROVE DATA AND REDUCE EMISSIONS*

United States Government Accountability Office

ABBREVIATIONS

BLM	Bureau of Land Management
DOE	Department of Energy
DOI	Department of the Interior
EIAEnergy	Information Administration
EPA	Environmental Protection Agency
FERC	Federal Energy Regulatory Commission
GGFR	Global Gas Flaring Reduction Partnership
IEA	International Energy Agency
LNG	Liquid Natural Gas
NOAA	National Oceanic and Atmospheric Administration
mmBtu	million British thermal units

* This is an edited, reformatted and augmented version of a report, GAO-04-809, issued by the U.S. Government Accountability Office, July 2004. Portions of the report have been omitted due to changes in the data since the report was originally issued. The full unedited report can be located on GAO's website (www.gao.gov).

MMS	Mineral Management Service
USAID	U.S. Agency for International Development

Why GAO Did This Study?

Natural gas is regularly lost as it is burned (flared) and released into the atmosphere (vented) during the production of oil and gas. GAO was asked to (1) describe flaring and venting data and what the federal government could do to improve them; (2) report, on the basis of available information, on the extent of flaring and venting and their contributions to greenhouse gases; and (3) identify opportunities for the federal government to reduce flaring and venting.

What GAO Recommends

GAO recommends that the Secretary of Energy consider opportunities to improve data on flaring and venting. In addition, GAO recommends that the Secretary of the Interior consider regulatory changes for federal leases to reduce the most harmful emissions from flaring and venting and to improve oversight.

What GAO Found

U.S. and global data on natural gas flaring and venting are limited. First, the Department of Energy's Energy Information Administration (EIA) collects and reports data voluntarily provided by oil- and gas-producing states. Because EIA has no authority to require states to report, some do not, leading to incomplete data. Second, EIA has provided limited guidance to states to promote consistent reporting. As a result, only about one-fourth of the states reporting provide data that EIA considers consistent. Third, the data EIA collects do not distinguish between flared gas and vented gas—an important distinction since they have dramatically different environmental impacts. Data on flaring and venting outside the United States are also limited, since many countries report unreliable data or none at all. To improve data on flaring and venting, EIA could use its authority to collect data directly from oil and gas producers; to obtain more consistent data, EIA could improve its guidelines for reporting. From an environmental perspective, EIA, the Minerals Management Service, and the Bureau of Land Management could require flaring and venting data to be reported separately from each other. Globally, the federal government could set an example by continuing to improve U.S. data, continuing to support global efforts, and using U.S. satellite data to detect unreported flaring.

On the basis of the limited data available, the amount of gas emitted through flaring and venting worldwide is small compared with global natural gas production and represents a small portion of greenhouse gas emissions. Nevertheless, flaring and venting have adverse environmental impacts and result in the loss of a significant amount of energy. Annually, over 100 billion cubic meters of gas are flared or vented worldwide—enough to meet the natural gas needs of France and Germany for a year. While flaring and venting do occur in the United States, less than 1 percent of global production is flared and vented.

Opportunities exist in several areas to help reduce flaring and venting, both in the United States and globally. For example, exploring ways to address market barriers affecting associated gas could help identify approaches to reduce global flaring and venting.

* * *

July 14, 2004

The Honorable Jeff Bingaman
Ranking Minority Member
Committee on Energy and Natural Resources
United States Senate

Dear Senator Bingaman:

Natural gas is regularly burned and released into the atmosphere during the production of oil and gas. While the exact amount of gas lost in this way is uncertain, given the vast extent of oil and gas production throughout the world, it could be significant. In addition to resulting in the loss of a potentially valuable resource, the burning and release of natural gas into the atmosphere contribute to greenhouse gas emissions, which are generally considered to be warming the earth's atmosphere.

Although most natural gas production involves extracting gas from wells drilled into underground gas reservoirs, some natural gas is generated as a by-product of oil production. Gas produced during oil production is called associated gas. During oil and gas production, it may be necessary to burn or release natural gas for a number of operational reasons, including lowering the pressure to ensure safety. Burning natural gas is known as flaring, while releasing natural gas directly into the atmosphere is called venting. In addition to the operational reasons for flaring and venting, in areas where the primary purpose of drilling is to produce oil, producers flare or vent associated natural gas because no local market exists for the gas and transporting it to a market may not be economically feasible.[1] Natural gas prices are a major determinant of whether associated gas is flared and vented or sold. Associated natural gas would be sold if prices were high enough over a long enough period to justify building the infrastructure—pipelines and ports—to transport the gas to a market. In the United States, there are well-developed natural gas markets and infrastructure to reduce the flaring and venting of associated natural gas. However, in parts of the world like Africa and the Middle East, where the natural gas market and infrastructure for transporting gas are not as well-developed, flaring and venting are generally more prevalent.

Although gas is sometimes flared or vented because it has too little economic value to justify its capture, flaring or venting this gas into the atmosphere has an environmental cost. In general, flaring emits carbon dioxide, while venting releases methane.[2] These and other chemical compounds found in the earth's atmosphere create a greenhouse effect. Under normal conditions, when sunlight strikes the earth's surface, some of it is reflected back towards space as infrared radiation or heat. Greenhouse gases such as carbon dioxide and methane impede this reflection by trapping heat in the atmosphere. Methane is 23 times more potent than carbon dioxide in its ability to warm the atmosphere. While these gases occur naturally on earth and are emitted into the atmosphere, the expanded industrialization of the world over the last 150 years has increased the amount of emissions from human activity (known as anthropogenic emissions) beyond the level that the earth's natural processes can handle. Scientists generally agree that these increased greenhouse gases are contributing to global warming, which can have detrimental effects on the climate. In general, the environmental costs of flaring and venting are not borne by the responsible parties because there are no restrictions on greenhouse gas emissions. However, nations have proposed international agreements to limit greenhouse gas emissions, including those released during flaring and venting.

Numerous federal agencies have a role in oil and gas production and its emissions' impact on the environment. The Department of Energy's (DOE) Energy Information Administration (EIA) is the federal agency responsible for providing energy information to meet the needs of the public, the federal government, industry, and the Congress for making policy, fostering efficient markets, and improving public understanding. DOE has also provided funding support to a World Bank public-private partnership to reduce global gas flaring. The Environmental Protection Agency (EPA) has responsibility for monitoring and reporting on a wide range of environmental issues, including greenhouse gases such as carbon dioxide and methane. The Department of the Interior's Minerals Management Service (MMS) and Bureau of Land Management (BLM) have responsibility for regulation and oversight of oil and gas leases for offshore federal properties and onshore federal lands, respectively. In general, flaring and venting are not authorized on these properties except for standard operational reasons. In addition, both agencies require producers to report combined flaring and venting data for federal leases. The Federal Energy Regulatory Commission (FERC) has responsibility for determining public convenience and necessity for interstate transmission of natural gas through pipelines as well as for permitting onshore liquefied natural gas facilities. In addition, the U.S. Coast Guard plays a role in the LNG offshore permitting process. The U.S. Agency for International Development (USAID) supports various projects around the world, including energy projects, in partnership with other countries. The National Oceanic and Atmospheric Administration (NOAA), under the Department of Commerce, uses U.S. satellite technology to, among other things, monitor gas flaring activity around the world.

Regarding natural gas flaring and venting from oil and gas production, in both the United States and the rest of the world, you asked us to (1) describe the data collected and reported on natural gas flaring and venting and what the federal government could do to improve it; (2) report, on the basis of available information, on the extent of flaring and venting and its contribution to greenhouse gases; and (3) identify opportunities for the federal government to reduce such flaring and venting.

To do this work, we obtained currently available data on natural gas production and estimates of flaring and venting in the United States from EIA, MMS, and BLM. We obtained international data from Cedigaz, a French oil and gas industry association that gathers worldwide information on natural gas; the World Bank; the International Energy Agency, an intergovernmental energy policy body; and the United Nations. We determined that all the data we reviewed were sufficiently reliable for inclusion in this report after acknowledging the limitations of these data. We also interviewed officials from EIA, EPA, MMS, BLM, NOAA, USAID, the World Bank, the United Nations, various private corporations and organizations, and state governments regarding data collection and methodologies, quality of the data collected, and reporting practices. In addition, we contacted natural gas-producing states to determine their assessment of the reliability of data they collect and report. We also attended the World Bank Global Gas Flaring Reduction Partnership's Second International Gas Flaring Reduction Conference in May 2004 in Algeria with delegates from numerous other countries concerned about gas flaring and venting. We conducted our work from October 2003 through June 2004 in accordance with generally accepted government auditing standards. A detailed description of our objectives, scope, and methodology is contained in appendix I.

RESULTS IN BRIEF

Overall, data collected and reported on the flaring and venting of natural gas associated with oil and gas production are limited in several ways. First, although the Department of Energy's EIA has authority to require U.S. oil- and gas-producing companies to report flaring and venting data, according to EIA officials, the agency has not used this authority because it considers these data a relatively low priority. Instead, EIA collects and publishes data provided voluntarily by the states on the amount of natural gas flared and vented. Because EIA has no authority to require states to report—and some do not—EIA data are incomplete and must be estimated by EIA. Compounding this problem, EIA has provided only limited guidance to states to promote consistency in the information that they voluntarily submit. As a result, only 8 of the 32 oil- and gas-producing states that EIA reports on provide data that EIA considers consistent, leaving the agency to estimate the amount of flaring and venting in the other 24 states. Finally, the information available to EIA does not distinguish between gas that is flared and gas that is vented, an important distinction from an environmental perspective because the methane emitted during venting is significantly more potent in its ability to warm the atmosphere than the carbon dioxide emitted during flaring.

Outside the United States, information on gas flared and vented is even more limited. Generally, international reporting is also voluntary. Moreover, there is no single organization responsible for collecting and reporting data.

The United Nations requests data as part of its work on climate change, but few countries report meaningful data. EIA reports worldwide as well as U.S. data, but it largely relies on estimates developed by an oil and gas industry association that gathers information self-reported by countries. In some cases, countries do not report the information or report suspect numbers. For example, Russia reports no flaring, even though satellite data analyzed by the National Oceanic and Atmospheric Administration have confirmed significant flaring activity. Also, as with the U.S. data, the global data do not distinguish between the amounts flared and vented. More accurate data would provide a clearer understanding of the extent and location of the problem, as well as a basis for targeted actions designed to both prevent the loss of a potentially valuable resource and reduce harmful emissions into the atmosphere.

The federal government could consider a number of opportunities to improve information on flaring and venting activity, if it deems better data is a worthwhile priority. EIA could improve its guidelines to states for collecting and reporting flaring and venting data, which could provide EIA with more consistent data. EIA could also consider collecting data on flaring and venting directly from the roughly 20,000 domestic oil and gas producers. In addition, from an environmental perspective, EIA, MMS, and BLM could require oil and gas producers to distinguish between the amount of gas flared and the amount vented. On a worldwide basis, the government could set an example for the world by continuing to improve U.S. data. In addition, EIA could work with NOAA to assess the feasibility of using U.S. satellite technology to quantify the volumes of gas flaring around the world and identify the additional resource commitment necessary to obtain the information each year. Finally, the federal government could continue to support global efforts such as the World Bank's Global Gas Flaring Reduction Partnership (GGFR), which, in part, seeks to develop better standards for data.

On the basis of the limited data available, the amount of gas emitted through flaring and venting is small compared with overall natural gas production and represents a small portion of greenhouse gas emissions. Nevertheless, flaring and venting do have adverse environmental impacts and result in the loss of a significant amount of energy. The amount of gas flared and vented annually is conservatively estimated by the World Bank at over 100 billion cubic meters worldwide, or about 3 percent of all gas marketed in the world. While this amount is small in comparison with worldwide totals, it is enough gas to meet the natural gas needs of both France and Germany for a year. Flaring and venting are concentrated in certain parts of the world. According to the World Bank, eight nations account for 60 percent of the estimated natural gas flared and vented: Algeria, Angola, Indonesia, Iran, Mexico, Nigeria, Russia, and Venezuela. Some countries flare and vent most of the associated natural gas they produce. In contrast, EIA estimates that the United States flares or vents about 0.4 percent of its production, representing only 3 percent of the world's total amount of natural gas flared and vented. Within the United States, most of the reported flaring and venting has occurred in the active oil and gas production states of Alaska, Louisiana, Texas, and Wyoming and in the Gulf of Mexico. Regarding worldwide greenhouse gas emissions, flaring and venting is estimated to contribute, respectively, about 4 percent of the total methane emissions and about 1 percent of the total carbon dioxide emissions caused by human activity. Several countries have taken steps to reduce flaring and venting emissions that have the potential of saving energy resources while reducing greenhouse gases. For example, some countries have imposed requirements on oil and gas producers to eliminate emissions of gas within the next few years.

The federal government has opportunities in several areas to help reduce natural gas flaring and venting. First, to address flaring and venting in the United States, MMS and BLM could consider regulatory changes that would reduce the most harmful emissions associated with flaring and venting and improve oversight of oil and gas production on federal lands and offshore areas leased to producers. Specifically, from an environmental perspective, it may be worth exploring the cost and benefit of requiring producers to flare rather than vent when they must release gas for standard operational purposes—for example, for safety reasons—since flaring is less detrimental to the atmosphere. In addition, while the identification of unauthorized flaring or venting is not commonplace on federal lands and offshore areas leased by oil and gas companies, unauthorized flaring and venting do occur. For example, a major oil and gas producer recently paid a $49 million settlement for unauthorized flaring and venting that went undetected for several years. Requiring the use of meters to measure gas flared and vented could improve oversight. Second, the federal government could promote programs that identify, and help industry implement, best practices for reducing natural gas emissions. For example, EPA sponsors the Natural Gas STAR program, which, among other things, identifies ways to reduce the need to flare and vent during oil and gas production operations. On a global basis, the U.S. government could investigate market barriers and the public's perceptions of the risks associated with new infrastructure, such as those related to building LNG facilities, which could affect relevant markets. For example, the permit approval process for building an LNG facility in the United States can currently take more than a year. Finally, the federal government could continue to work with other countries and international corporations to reduce flaring and venting.

We are making recommendations to the Secretary of Energy to consider the opportunities to improve data on flaring and venting. We are also making recommendations to the Secretary of the Interior to consider regulatory changes to reduce the most harmful emissions from flaring and venting and to improve oversight for federal leases.

FLARING AND VENTING DATA ARE LIMITED, BUT THE FEDERAL GOVERNMENT HAS OPPORTUNITIES TO IMPROVE THEM

In the United States, DOE's EIA collects and reports data provided voluntarily by the states on the amount of natural gas flared and vented, but the data are incomplete, inconsistent, and not as useful as they could be from an environmental perspective. Information on gas flared and vented outside the United States is also limited, since international reporting generally is voluntary and there is no single organization that is responsible for collecting and reporting this data. EIA could improve flaring and venting data by enhancing its guidance to states and collecting data directly from oil and gas producers; EIA, MMS, and BLM could also improve data by collecting and reporting data on venting separately from data on flaring. By taking these actions, the federal government could serve as a model for global data collection and reporting. EIA could also investigate improvements in global data collection and reporting by using satellite images and data to better estimate the volume of natural gas flared in other countries and by continuing to support international efforts to improve data.

Data Collected and Reported on Flaring and Venting Are Limited

In the United States, data collected and reported on the flaring and venting of natural gas associated with oil and gas production are incomplete, inconsistent, and not as useful as they could be from an environmental perspective. Regarding the completeness of the data, although MMS and BLM require companies that lease federal lands and offshore areas for oil and gas production to report flaring and venting statistics, EIA does not use its authority to require information on flaring and venting from all other U.S. oil- and gas-producing companies. According to EIA officials, the data on flaring and venting that EIA collects as part of overall oil and gas production data represent a relatively small portion of EIA's energy data reporting program. Since the agency has limited resources, these data are a relatively low priority. As a result, rather than requiring that the estimated 20,000 domestic oil- and gas-producing companies provide flaring and venting data—which would consume considerable resources—EIA collects this information from the oil- and gas-producing states on a voluntary basis. Many states do not provide this information, however, and EIA has no authority to require them to do so. Consequently, EIA's flaring and venting information is incomplete.

In addition to being incomplete, the data the states provide to EIA are also inconsistent. Since flaring and venting data are not a high priority for EIA, the agency has provided only limited guidelines to states to promote consistency in the information that they voluntarily submit. As a result, only 8 of the 32 oil- and gas-producing states [3] provide data that EIA considers consistent, leaving EIA to estimate the amount of flaring and venting in the other 24 states. When we asked state officials about EIA guidelines for reporting, officials from 15 states said they were unsure what information EIA wanted and how EIA wanted it presented. Most officials from the states answering our questions about the guidelines thought they needed improvement, and some officials said they would like to participate in developing improved guidelines to ensure that the states would be able to meet EIA's requests.

The data that EIA, MMS, and BLM collect are further limited because they do not distinguish between gas that is flared and gas that is vented. As a result, from an environmental perspective, the information is not as useful as it could be. EIA, MMS, and BLM do not collect separate flaring and venting data because their focus is on the amount of gas produced for the market and not on the gas that is lost through flaring and venting. EPA, on the other hand, considers flared and vented gas in the context of the damage it could inflict on the environment: vented gas emissions (methane), and to a lesser extent flared gas emissions (carbon dioxide), contribute to total greenhouse gases. According to EPA officials, differentiating data on flaring and venting could improve EPA estimates of each gas's contribution to total greenhouse gases in the atmosphere. Because EPA does not collect its own flaring and venting data, however, it must rely on the combined data that these other agencies collect.

EIA believes that the data the states voluntarily provide on production—from which the data on flaring and venting are taken—could be improved as well. In particular, because some states do not report information at all, EIA is considering using its authority to collect information on production directly from natural gas well operators. Toward that end, EIA has published for comment a proposed sample survey of monthly natural gas production in the Federal Register. If the survey were implemented, well operators would be required to provide EIA with production data and the form would include a category for flaring and venting data.[4] Among other things, EIA has sought comments from well operators as to whether they can provide reliable measures of gas flared and vented. However, even if the proposed sample form is implemented and, as proposed, collects information on flared and vented data, the focus will be on improving production data and not on flaring and venting data.

Outside the United States, information on gas flared and vented is even more limited. Generally, international reporting is also voluntary, and no single organization is responsible for collecting and reporting flaring and venting data. Although several organizations collect data voluntarily provided by countries with which they have a working relationship, the numbers countries report are sometimes questionable. For example, the United Nations requests data as part of its work on climate change, but few countries report meaningful data. In addition to U.S. data, EIA also reports worldwide information largely based on estimates developed by Cedigaz, an oil and gas industry association that gathers what is generally recognized as the best flaring and venting information available. A Cedigaz official told us that they rely on submissions from countries and companies around the world to make their estimates and that they accept the information the countries report unless Cedigaz has knowledge of a country's operations that could be used to improve the accuracy of the amounts reported. For example, on the basis of submissions by Russia and China—two important petroleum-producing countries—Cedigaz has reported that these countries do not flare or vent. World Bank officials told us, however, it is generally known that Russia regularly flares and vents gas. Satellite images created by NOAA have confirmed that Russia does, in fact, participate in flaring. In addition, as in the United States, the global data are limited from an environmental perspective because they do not distinguish between the amounts flared and the amounts vented. More accurate worldwide data would provide a clearer understanding of both the extent to which flaring and venting emissions contribute to total greenhouse gases and the countries that do the most flaring and venting. This would provide a basis for targeting actions designed to prevent the waste of a potentially valuable resource while at the same time reducing harmful emissions into the atmosphere.

The Federal Government Has Several Opportunities to Improve the Information on Flaring and Venting

Federal agencies have a number of opportunities available to them to improve the information on flaring and venting. EIA could clarify its guidelines to states for collecting and reporting flaring and venting data. Currently, EIA assumes that about 75 percent of the reports it receives from oil- and gas-producing states contain inconsistent data. State officials believe that they could better meet EIA's data needs—that is, provide EIA with more consistent data—if they had more comprehensive guidance from EIA on the data it wants and how to report them. Similarly, as the reporting of greenhouse gas emissions has become more widespread globally, the oil and gas industry has begun to recognize the need for guidance on how emissions, such as carbon dioxide and methane, should be collected and reported. In December 2003, three petroleum industry associations jointly issued a report, "Petroleum Industry Guidelines for Reporting Greenhouse Gas Emissions," to promote consistency in collecting and reporting petroleum industry greenhouse gas emissions. EIA could consider using these guidelines while working with industry and state officials to improve their state guidelines for reporting emissions from flaring and venting of natural gas.

Another opportunity for improving the data is for EIA to consider using its general energy information collecting authority [5] to collect data on flaring and venting directly from the oil and gas producers, rather than relying on voluntary submissions by states. (Producers with federal leases are already required to collect and report this information to MMS and BLM.) While EIA considers these data a relatively low priority, and while collecting data from all 20,000 domestic producers could involve extensive resources, there may be efficient and cost-effective methods of collecting sample data. In addition, from an environmental perspective, the federal government could broaden the usefulness of flaring and venting information by distinguishing between the amounts of gas flared and the amounts vented. Since natural gas that is vented has a more significant effect on the atmosphere than natural gas that is flared, reporting these emissions separately could enable EPA to better estimate methane's and carbon dioxide's contributions to greenhouse gases.

Finally, on a worldwide basis, the federal government could improve flaring and venting data in several ways. First, the U.S. government could continue to improve its own data, thereby providing an example for other countries. Second, EIA, working with NOAA, could investigate the feasibility of supplementing the data already available by using U.S. satellite images. According to a NOAA physical scientist, analyzing these satellite data could validate the amount of flaring reported by countries. For example, some countries, like Russia, report no flaring, while satellite images show substantial flaring is actually occurring. Third, the federal government could continue to support efforts such as the World Bank Global Gas Flaring Reduction Partnership (GGFR)[6] that, in part, seeks to develop guidance for data collection and reporting. According to a 2004 GGFR report, such guidance could improve natural gas flaring and venting data.

Although the Amount of Gas Flared and Vented Is Small, It Represents a Significant Amount of Lost Energy and Contributes to Global Greenhouse Gas Emissions

According to the limited data available, the amount of natural gas emitted through flaring and venting is small compared with overall natural gas production, but these emissions represent a significant amount of lost energy. Flaring and venting are concentrated in several parts of the world and, in the United States, in four states and the Gulf of Mexico. Although worldwide estimates of flaring and venting constitute a small portion of total greenhouse gas emissions, many countries have undertaken efforts to reduce flaring and venting.

Three Percent of the World Production of Natural Gas Is Flared or Vented, Which Represents a Significant Amount of Unused Resources

While flaring and venting represent only 3 percent of the total natural gas production, the natural gas flared and vented—about 100 billion cubic meters a year—is enough to meet the annual natural gas consumption of both France and Germany. In general, the amount of flaring and venting emissions is related to the amount of oil produced: the higher the production, the more gas flared and vented.

Natural Gas Flaring and Venting Emissions Contribute a Small Percentage of Global Greenhouse Gas Emissions, and Countries Have Undertaken Initiatives to Reduce These Emissions

Worldwide flaring and venting is estimated to contribute, respectively, about 4 percent of the total methane and about 1 percent of the total carbon dioxide emissions caused by human activity. Despite these small contributions, several countries have undertaken efforts to reduce flaring and venting emissions that have the potential to reduce greenhouse gases while saving an energy resource. Specifically, many countries have imposed requirements on oil and gas producers to eliminate emissions of gas within the next few years. For example, Norway no longer allows the burning of petroleum in excess of the quantity needed for normal operational safety without the approval of the Ministry of Petroleum and Energy, and in 2003 Canada reported having achieved, through monitoring and regulation, a 70 percent reduction in flaring and venting emissions. In addition, corporations in several countries, in order to market their associated natural gas, either have constructed or are planning LNG plants to liquefy the gas for export or have developed on-site and local uses for the gas. For example, corporations operating in Nigeria currently have six LNG projects in development and have also begun using gas that otherwise would have been flared or vented to operate the platform equipment as well as to produce cement and fertilizer and gas that is usable as fuel for automobiles. Finally, some countries are exploring the potential of reinjecting carbon dioxide into wells instead of emitting it into the atmosphere. According to an oil company official, carbon dioxide reinjection in Algeria has prevented over one million tons of emissions— the equivalent of taking 200,000 cars off the road.

The Federal Government Could Take a Larger Role in Reducing Flaring and Venting

The federal government has opportunities in several areas to help reduce flaring and venting both within the United States and globally. First, in the

United States, on federal lands and offshore areas leased to producers, the federal government could consider regulatory changes to reduce the most harmful emissions resulting from venting and improve oversight of oil and gas production. Second, the federal government could promote programs that identify, and help industry implement, best practices for reducing natural gas emissions.

On a global basis, exploring ways to address market barriers affecting associated natural gas, and continuing to work with other countries, could help reduce flaring and venting.

Improving Federal Regulation and Oversight Could Reduce Flaring and Venting on Federal Lands

The government has an opportunity to help reduce flaring and venting in the United States by considering regulatory changes and improving oversight of oil and gas production on federal lands and offshore areas leased to producers. If MMS and BLM required producers to flare gas (rather than allowing them to vent gas) when emitting the gas for operational purposes, the emissions impact on the atmosphere could be reduced. Since the impact of methane (venting) on the earth's atmosphere is about 23 times greater than that of carbon dioxide (flaring), a small change in the ratio of flaring to venting could cause a disproportionate change in the impact of emissions. For example, if 90 percent of the associated gas volume was flared and 10 percent was vented, the amount vented would have more than twice the effect on the atmosphere as the amount flared.

In addition, although MMS and BLM require federal lessees to self-report estimated flaring and venting volumes, the agencies could require the use of flare and vent meters at production facilities, which could improve oversight by detecting how much gas is actually flared and vented. Currently, such meters are used by some oil- and gas-producing companies. The meters are placed in or adjacent to the stream of flared or vented gas to measure the volume emitted. While the identification of unauthorized flaring or venting is not commonplace on federal lands and offshore leases in the United States, unauthorized flaring and venting do occur. A major oil and gas producer recently paid a $49 million settlement in response to charges of unauthorized flaring and venting in the Gulf of Mexico that went undetected for several years.

The lawsuit alleged that the company both improperly flared and vented gas under various offshore federal leases and failed to properly report the flared and vented gas. Improved oversight could help prevent similar problems in the future. The International Energy Agency has recommended such meters for accurately monitoring emissions from production facilities, and some producers that we interviewed said they already use these meters on many of their offshore production facilities. According to these producers, flare and vent meters are fairly inexpensive if installed when the facility is initially built, but retrofitting facilities that are already producing may not be cost effective.[7] To improve oversight, MMS and BLM, working with producers, could consider the cost and benefit of requiring the installation of these meters, particularly at new, major production facilities.

Promoting Best Practices Could Reduce Emissions Caused by Flaring and Venting

The federal government could also help reduce flaring and venting by continuing to support programs that identify, and help industry implement, best practices for reducing natural gas emissions. EPA has sponsored the Natural Gas STAR program, which identifies and promotes the implementation of cost-effective technologies and practices that reduce methane emissions. About 65 percent of the U.S. natural gas industry is represented in this program. From 1993 through January 2004, according to EPA officials the STAR program identified over 100 best practices and technologies that reduce methane emissions from the oil and natural gas industry.

In addition, STAR participants have reported total emissions reductions of more than 275 billion cubic feet worth over $825 million— which EPA estimates is enough natural gas to heat more than 4 million homes for 1 year or is comparable to the removal of the emissions of 24 million cars from the nation's highways for 1 year. The two largest sources of overall oil and gas emissions identified by the STAR program are pneumatic devices and compressors.

According to EPA officials, the STAR Program has identified 15 practices and technologies that would reduce methane emissions from these sources.

Investigating Market Barriers Affecting Associated Gas and Continuing to Work with Other Countries Could Help Reduce Global Flaring and Venting

The government could also have an effect on flaring and venting worldwide by (1) addressing market barriers affecting gas produced outside the United States that would otherwise be flared or vented and (2) continuing to work with other countries. First, the government could identify regulatory barriers to economically feasible infrastructure development, such as building pipelines or LNG facilities for transporting gas that is usually flared or vented. For example, currently, four LNG import facilities exist in the United States—three on the East Coast and one on the Gulf of Mexico coast—but more than 15 federal permit applications are awaiting approval (from either FERC or the Coast Guard), which can take more than a year.[8] In addition, the government could investigate the public's perceptions of the risk associated with new infrastructure. For example, some communities have resisted LNG facilities because they are worried about the safety and security procedures in place to protect them from an accidental explosion or a terrorist attack. Finally, the federal government could continue to work with other countries and corporations to reduce flaring and venting. For example, USAID provided much of the funding for training regulators in Kazakhstan, where improved regulation has virtually eliminated the routine flaring of natural gas. In addition, the United States could continue to support the work of the World Bank's Global Gas Flaring Reduction Partnership (GGFR), which recently issued standards on how to achieve reductions in the flaring and venting of gas worldwide.[9]

CONCLUSION

By itself, the reduction of natural gas flaring and venting will not solve the problem of meeting increasing natural gas demands or eliminate all greenhouse gas emissions; however, it would be a helpful step in that direction. Although the emissions from flaring and venting are small in comparison with those from other sources, such as fossil fuel combustion, reducing flaring and venting from oil and gas production would help eliminate harmful emissions and possibly preserve an energy resource that is currently being lost.

Although the role of the federal government in reducing flaring and venting of natural gas during oil and gas production may be limited, especially at the international level, opportunities exist for the government to make worthwhile contributions in this area. Given the immense challenges the government faces in the energy and environmental arenas and the limited resources available, however, any actions to reduce flaring and venting will have to be based on careful consideration of the potential costs and benefits of such actions. Moreover, since flaring and venting are not viewed as major problems in this country, it may be difficult to justify devoting much attention to them. Nonetheless, the government could consider several potentially cost-effective actions that would improve data reporting, help reduce the most harmful effects of flaring and venting, and improve oversight of federal leases.

Improving flaring and venting data reporting would be an important first step because it would add information about the scope of the problem and allow actions to be targeted to those areas where flaring and venting are most significant. However, our review identified several limitations of the data currently reported that hinder such targeting. First, reliance on oil- and gas-producing states to voluntarily provide flaring and venting information has led to incomplete data, since some states do not submit reports. Second, because of the limited guidance on what data to report, EIA considers most of the information it receives from the states to be inconsistent. Third, the data collected do not distinguish between flared gas and vented gas—an important distinction, since the methane emitted during venting is potentially much more harmful to the atmosphere than the carbon dioxide emitted during flaring. Consequently, these combined data are less useful than they could be in shedding light on the extent of methane emitted into the atmosphere during venting. Finally, on a global basis, some countries that flare and vent considerable amounts of gas report that they do not, and others underreport their flaring and venting. For these and other reasons, the worldwide data are considered even less accurate than U.S. data.

In addition, for federal lands and offshore areas leased for oil or gas production, opportunities exist to help reduce the most harmful effects of flaring and venting and improve oversight on these leases. Specifically, although flaring and venting are authorized only under certain circumstances, when these circumstances occur, oil and gas producers may choose to either flare or vent. Since the emissions released during venting are much more harmful to the atmosphere than those associated with flaring, this choice is an important one. Moreover, although flaring and venting are generally not authorized, no oversight mechanism currently exists for routinely monitoring the amount of flaring and venting that actually takes place. As a result, MMS and BLM cannot always be assured that companies are appropriately restricting their flaring and venting.

Overall, however, as is evident in the example of the United States, a robust natural gas market, along with a supporting infrastructure, would have the most significant impact on the reduction of flaring and venting. Therefore, changes to natural gas markets, and to the transportation infrastructure for moving gas to these markets, will likely be needed to offer producers an economic incentive to sell the associated gas rather than flare or vent it.

RECOMMENDATIONS FOR EXECUTIVE ACTION

We are making four recommendations to the Secretary of Energy to explore opportunities for improving data on flaring and venting and to weigh the cost and benefit of making such improvements. Such opportunities could include

- considering the use of the department's authority to collect flaring and venting information directly from the producing companies;
- working with industry and state officials, and using guidance already in existence, to enhance guidelines for reporting consistent flaring and venting data;
- considering, in consultation with EPA, MMS, and BLM, how to best collect separate statistics on flaring and venting; and
- considering working with the Secretary of Commerce to provide EIA access to NOAA's analysis of satellite data to improve the accuracy of worldwide data on flaring.

We are also making two recommendations to the Secretary of the Interior. Specifically, for federal oil and gas leases, we are recommending that the Secretary direct MMS and BLM to consider the cost and benefit of requiring that companies

- flare the natural gas, whenever possible, when flaring or venting is necessary and
- use flaring and venting meters to improve oversight.

AGENCY COMMENTS AND OUR EVALUATION

We requested comments on a draft of this report from the Secretaries of Energy and the Interior, as well as the Administrator of the Environmental Protection Agency. The Department of Energy and Department of the Interior provided written responses, including technical and clarifying comments, which we included in our report as appropriate.

The Administrator of the Energy Information Administration (EIA), who responded on behalf of the Department of Energy, concurred with our findings and recommendations while acknowledging that the implementation of the recommendations would need to be balanced with other priorities. EIA stated it has revised the instructions on all natural gas forms and has committed to continue efforts to improve data quality, timeliness and coverage of production data that states collect. Further, EIA agreed to work with agencies with satellite technology to determine the feasibility of improving data worldwide.

The Assistant Secretary of Land and Minerals Management, who responded on behalf of the MMS and BLM within the Department of the Interior, generally agreed with our findings and recommendations. MMS and BLM agreed that requiring operators to flare instead of vent whenever possible would reduce greenhouse gases. Further, MMS acknowledged that this would enhance the agency's ability to enforce existing regulations because inspectors could easily view a flare; whereas inspectors cannot be detected venting visually and thus they must rely on the accuracy of operators' records. The two agencies also agreed to evaluate the cost effectiveness of installing and maintaining flare tips. In addition, MMS agreed that requiring flaring and venting meters would improve its oversight. MMS acknowledged that recent incidents have shown that reliance on the accuracy of the operators' calculations and recordkeeping may not sufficiently or accurately capture actual flaring and venting volumes. MMS stated that it is currently revising its Federal Outer Continental Shelf regulations to require installation of meters. Finally, MMS agreed that flaring and venting data should be reported separately and has taken initial steps to update its database and reporting requirements to accommodate this change. However, MMS stated that such a change would take time to complete and would only minimally add to the reporting burden of operators. The Department of the Interior's comments are reprinted in appendix III.

EPA provided oral comments and agreed with our findings and recommendations. In addition, EPA noted that our recommendations, if implemented, would greatly advance flaring and venting data availability and quality. EPA also provided technical comments, which we included in the report, as appropriate.

Sincerely yours,
Jim Wells
Director,
Natural Resources and Environment

OBJECTIVES, SCOPE, AND METHODOLOGY

Regarding natural gas flaring and venting from oil and gas production in the United States and the rest of the world, you asked us to (1) describe the data collected and reported on natural gas flaring and venting and what the federal government could do to improve it; (2) report, on the basis of available information, on the extent of flaring and venting and its contribution to greenhouse gases; and (3) identify opportunities for the federal government to reduce such flaring and venting.

To do this work, we obtained currently available data on natural gas production and estimates of flaring and venting in the United States from EIA, MMS, and BLM. We determined that all the data we reviewed were sufficiently reliable for inclusion in this report after acknowledging the limitations of these data. We also interviewed officials from EIA, EPA, MMS, BLM, the World Bank, the United Nations, various private corporations and organizations, and state governments regarding data collection, quality of the data collected, and reporting practices. In addition, we contacted natural gas-producing states to determine their assessment of the reliability of data they collect and report. To determine what the federal government could do to improve data collecting and reporting, we interviewed officials from EIA, MMS, BLM, EPA, NOAA, and state officials. In addition, we obtained international data from Cedigaz, a French oil and gas industry association that gathers worldwide information on natural gas; the World Bank; the International Energy Agency (IEA), an intergovernmental energy policy body; and the United Nations. We discussed with officials of these organizations the reliability as well as the methods of collecting and estimating these data.

Finally, to determine what the federal government could do to help reduce flaring and venting of natural gas, we reviewed the literature and interviewed officials from private industry, MMS, BLM, state governments, EPA's Natural Gas STAR program, USAID, and various world organizations, including the

World Bank, the United Nations, and IEA. In May 2004, we also attended the World Bank Global Gas Flaring Reduction Partnership's Second International Gas Flaring Reduction Conference in Algeria with delegates from numerous other countries concerned about gas flaring and venting. This conference featured speakers from countries and corporations around the world who addressed data collection and reporting issues, current regulations and oversight issues, and best practices within their respective countries. We conducted our work from October 2003 through June 2004 in accordance with generally accepted government auditing standards.

End Notes

[1] Natural gas may sometimes be reinjected into the ground for possible future use as an alternative to flaring or venting. Reinjection can be more costly than flaring and venting, but in some cases it may enhance oil recovery by increasing pressure in the reservoir.

[2] Besides carbon dioxide, flaring natural gas may emit small amounts of methane and trace amounts of other gases depending on the flare's burn efficiency. Venting methane may also release other hydrocarbons depending on the composition of the gas stream. In addition, due to the composition of the gas--for example when gas contains hydrogen sulfide--flaring may be required to control any associated air quality or human health impacts.

[3] EIA does not report information for a 33rd state, Missouri, which produces a small amount of oil and gas.

[4] 69 Fed. Reg. 22109 (Apr. 23, 2004).

[5] 15 U.S.C. § 772(b).

[6] GGFR is a public-private partnership that in May 2004 included BP, ChevronTexaco, ENI, ExxonMobil, NorskHydro, Royal Dutch Shell, SNH, Statoil, TOTAL, and Sonatrach; and the governments of Angola, Canada, Chad, Ecuador, Equatorial Guinea, Indonesia, Nigeria, Norway, and the United States.

[7] For example, industry officials told us that some of the older platforms have limited space, so to add meters would require additional construction, which could be cost prohibitive.

[8] Three additional facilities were recently approved but have not yet been built. Eleven other projects are in various planning stages. According to DOE officials, streamlined approval procedures require offshore deepwater port terminals to be processed in less than a year by the U.S. Coast Guard and onshore facility permitting is being streamlined by FERC through pre-filing consultations with project proponents to minimize approval time.

[9] A Voluntary Standard for Global Gas Flaring and Venting Reduction (World Bank: May 2004).

In: Flaring and Venting of Natural Gas
Editor: Linda R. Henderson
ISBN: 978-1-63463-838-8

Chapter 2

FEDERAL OIL AND GAS LEASES: OPPORTUNITIES EXIST TO CAPTURE VENTED AND FLARED NATURAL GAS, WHICH WOULD INCREASE ROYALTY PAYMENTS AND REDUCE GREENHOUSE GASES*

United States Government Accountability Office Report to Congressional Requesters

WHY GAO DID THIS STUDY

The Department of the Interior (Interior) leases public lands for oil and natural gas development, which generated about $9 billion in royalties in 2009. Some gas produced on these leases cannot be easily captured and is released (vented) directly to the atmosphere or is burned (flared). This vented and flared gas represents potential lost royalties for Interior and contributes to greenhouse gas emissions.

GAO was asked to (1) examine available estimates of the vented and flared natural gas on federal leases, (2) estimate the potential to capture

* This is an edited, reformatted and augmented version of a United States Government Accountability Office publication, GAO-11-34, dated October 2010.

additional gas with available technologies and associated potential increases in royalty payments and decreases in greenhouse gas emissions, and (3) assess the federal role in reducing venting and flaring. In addressing these objectives, GAO analyzed data from Interior, the Environmental Protection Agency (EPA), and others and interviewed agency and industry officials.

What GAO Recommends

To reduce lost gas, increase royalties, and reduce greenhouse gas emissions, GAO recommends that Interior improve its venting and flaring data and address limitations in its regulations and guidance. Interior generally concurred with these recommendations.

What GAO Found

Estimates of vented and flared natural gas for federal leases vary considerably, and GAO found that data collected by Interior to track venting and flaring on federal leases likely underestimate venting and flaring because they do not account for all sources of lost gas. For onshore federal leases, operators reported to Interior that about 0.13 percent of produced gas was vented or flared. Estimates from EPA and the Western Regional Air Partnership (WRAP) showed volumes as high as 30 times higher. Similarly, for offshore federal leases, operators reported that 0.5 percent of the natural gas produced was vented and flared, while data from an Interior offshore air quality study showed that volume to be about 1.4 percent, and estimates from EPA showed it to be about 2.3 percent. GAO found that the volumes operators reported to Interior do not fully account for some ongoing losses such as the emissions from gas dehydration equipment or from thousands of valves—key sources in the EPA, WRAP, and Interior offshore air quality studies.

Data from EPA, supported by information obtained from technology vendors and GAO analysis, suggest that around 40 percent of natural gas estimated to be vented and flared on onshore federal leases could be economically captured with currently available control technologies. According to GAO analysis, such reductions could increase federal royalty payments by about $23 million annually and reduce greenhouse gas emissions by an amount equivalent to about 16.5 million metric tons of CO_2—the annual emissions equivalent of 3.1 million cars. Venting and flaring reductions are also possible offshore, but data were not available for GAO to develop a complete estimate.

Source: EPA.

Vented Gas from Oil Storage Tank Visible through Infrared Camera.

As part of its oversight responsibilities, Interior is charged with minimizing vented and flared gas on federal leases. To minimize lost gas, Interior has issued regulations and guidance that limit venting and flaring during routine procedures. However, Interior's oversight efforts to minimize these losses have several limitations, including that its regulations and guidance do not address some significant sources of lost gas, despite available control technologies to potentially reduce them. Although EPA does not have a role in managing federal leases, it has voluntarily collaborated with the oil and gas industry through its Natural Gas STAR program, which encourages oil and gas producers to use gas saving technology, and through which operators reported venting reductions totaling about 0.4 percent of natural gas production in 2008.

ABBREVIATIONS

Bcf billion cubic feet
BLM Bureau of Land Management
BOEMRE Bureau of Ocean Energy Management, Regulation and Enforcement
EIA Energy Information Administration
EPA Environmental Protection Agency

GOADS	Gulfwide Offshore Activities Data System
Interior	Department of the Interior
MRM	Minerals Revenue Management
NTL	Notice to Lessees and Operators
OGOR	Oil and Gas Operations Report
WRAP	Western Regional Air Partnership
VOC	volatile organic compound

* * *

October 29, 2010

The Honorable Darrell Issa
Ranking Member
Committee on Oversight
and Government Reform
House of Representatives

The Honorable Nick J. Rahall, II
Chairman
Committee on Natural Resources
House of Representatives

Production of oil and natural gas on federal lands and waters is an important part of the nation's energy portfolio and a significant source of revenue for the federal government. The Department of the Interior (Interior) manages lands that account for nearly a quarter of domestic oil and gas production. In fiscal year 2009, companies that leased these lands paid about $6 billion in royalties to the federal government on the sale of oil and gas produced offshore in federal waters, and about $3 billion for production on federal lands, making revenues from federal oil and gas one of the largest nontax sources of federal government funds. Interior's Bureau of Land Management (BLM) is responsible for managing leases onshore, and its Bureau of Ocean Energy Management, Regulation and Enforcement (BOEMRE) is responsible for leases offshore.[1]

While most of the natural gas produced on leased federal lands and waters is sold, some is lost during production for various reasons, including leaks and releases for ongoing operational or safety procedures. This natural gas is either released directly into the atmosphere (vented) or burned (flared).[2] The venting and flaring of natural gas is the potential loss of a valuable resource and, on leased federal lands or waters, the loss of federal royalty payments. In addition, venting releases methane, and flaring emits carbon dioxide (CO_2), both greenhouse gases that contribute to global climate change. Methane is a particular concern since it is a more potent greenhouse gas than is CO_2.[3]

In 2004, we reported that Interior, the department charged with managing federal oil and gas leases and regulating venting and flaring, collected and reported information on the extent of venting and flaring on leased federal lands and waters.[4] We made two recommendations to Interior to improve the measurement of vented and flared gas and to reduce its impact, which the department implemented. Since that time, the Environmental Protection Agency (EPA) and the oil and gas industry identified sources of vented and flared gas that were releasing substantially more gas than previously thought possible, suggesting that the expanded use of available technologies could help capture additional gas. This report responds to your request that we review the extent of venting and flaring of natural gas on federal leases. Our objectives were to (1) examine available estimates of vented and flared natural gas on federal leases; (2) estimate the potential to capture additional vented and flared natural gas with available technologies and the associated potential increases in royalty payments and reductions in greenhouse gas emissions and; (3) assess the federal role in reducing venting and flaring of natural gas.

To examine estimates of the volumes of vented and flared natural gas on federal leases, we analyzed data on venting and flaring that oil and gas producers submit to Interior's Minerals Revenue Management (MRM) program, which is responsible for collecting revenue from federal leases. MRM uses these data from its Oil and Gas Operations Report (OGOR) data system to account for monthly oil and gas production onshore and offshore. Separate regulations and guidance from BLM and BOEMRE guide operators in reporting to OGOR, and these data are the primary information source these agencies use to monitor overall venting and flaring. We also analyzed data from BOEMRE's Gulfwide Offshore Activity Data System (GOADS), which BOEMRE collects and publishes in a study every 3 years.[5] BOEMRE uses its GOADS studies to estimate the impacts of offshore oil and gas exploration, development, and production on onshore air quality in the Gulf of Mexico region, which made up about 98 percent of federal offshore gas production in 2008. BOEMRE also uses GOADS as part of an impact analysis required by the National Environmental Policy Act. In addition, we analyzed EPA estimates of vented and flared gas onshore and offshore.[6] We also analyzed data on vented and flared natural gas that the Western Regional Air

Partnership (WRAP),[7] in conjunction with the Independent Petroleum Association of Mountain States,[8] collected from the oil and gas industry to measure air quality in a number of large production basins in the mountain west.[9] We had consultants from the Environ International Corporation, the firm that collected and analyzed the air quality data for WRAP, reconfigure these data to provide information on venting and flaring volumes on federal leases for a number of these onshore basins. Our sources of venting and flaring data were from 2006 to 2008, and we examined only the portions of these data related to federal leases in order to ensure comparability between them. We assessed the reliability of the data we used by analyzing the methods used to construct them and found them sufficiently reliable for the purposes of this report.

To estimate the potential federal royalty increases and greenhouse gas reductions resulting from capturing additional vented and flared gas with available technologies, we met and spoke with officials from the oil and gas industry and to vendors of products designed to reduce vented and flared gas about how and under what conditions these technologies could reduce venting and flaring. We used analyses and data from EPA and WRAP to estimate potential reductions in volumes of vented and flared natural gas on federal leases, then converted these volumes into potential federal royalty increases and greenhouse gas reductions using methane to carbon dioxide equivalent conversion factors, average natural gas prices, and royalty rates from Interior. To assess the federal role in reducing vented and flared gas, we interviewed officials from BLM and BOEMRE, including officials from field offices that manage oil and gas leases in large onshore and offshore production basins; EPA; the Department of Energy; state agencies; and the oil and gas industry. We also reviewed BLM and BOEMRE regulations and other documentation, other studies related to federal management and oversight of the oil and gas industry, as well as a prior GAO report that described limitations in the systems Interior has in place to track oil and gas production on federal leases.[10] See appendix I for more detailed information on our scope and methodology.

We conducted this performance audit from July 2009 to October 2010 in accordance with generally accepted government auditing standards. Those standards require that we plan and perform the audit to obtain sufficient, appropriate evidence to provide a reasonable basis for our findings and conclusions based on our audit objectives. We believe that the evidence obtained provides a reasonable basis for our findings and conclusions based on our audit objectives.

BACKGROUND

The Mineral Leasing Act of 1920 charges Interior with overseeing oil and gas leasing on federal lands and private lands where the federal government has retained mineral rights covering about 700 million onshore acres.[11] Offshore, the Outer Continental Shelf Lands Act,[12] as amended, gives Interior the responsibility for leasing and managing approximately 1.76 billion acres. BLM and BOEMRE are responsible for issuing permits for oil and gas drilling; establishing guidelines for measuring oil and gas production; conducting production inspections; and generally providing oversight for ensuring that oil and gas companies comply with applicable laws, regulations, and department policies. This oversight includes the authority to ensure that firms produce oil and gas in a manner that minimizes any waste of these resources. Together, BLM and BOEMRE are responsible for oversight of oil and gas operations on more than 28,000 producible leases.

Interior's MRM program, which is managed under BOEMRE, is charged with ensuring that the federal government receives royalties from the operators that produce oil and gas from both onshore and offshore federal leases. MRM is responsible for collecting royalties on all of the oil and gas produced, with some allowances for gas lost during production. Companies pay royalties to MRM based on a percentage of the cash value of the oil and gas produced and sold. Currently, royalty rates for onshore leases are generally 12.5 percent, while rates for offshore leases range from 12.5 percent to 18.75 percent.

The production of oil and gas on these federal leases involves several stages, including the initial drilling of the well; clearing out liquid and mud from the wellbore; production of oil and gas from the well; separation of oil, gas, and other liquids; transfer of oil and gas to storage tanks; and distribution to central processing facilities. Throughout this process, operators typically vent or flare some natural gas, often intermittently in response to maintenance needs or equipment failures. This intermittent venting may take place when operators purge water or hydrocarbon liquids that collect in well bores (liquid unloading) to maintain proper well function or when they expel liquids and mud with pressurized natural gas after drilling during the well completion process. BLM and BOEMRE permit operators of wells to release routine amounts of gas during the course of production without notifying them or incurring royalties on this gas.[13] In addition, production equipment often emits gas to maintain proper internal pressure, or in some cases, the release of pressurized gas itself is the power source for the equipment, particularly in remote areas that are not linked to an electrical grid. This "operational" venting may include the continuous releases of gas from pneumatic devices—valves that control gas flows, levels, temperatures, and pressures in the equipment and rely on pressurized gas for operation—as well as leaks, or "fugitive" emissions.[14] It also includes natural gas that vaporizes from oil or condensate storage tanks or during the normal operation of natural gas dehydration equipment.[15] Until recently, the industry considered these operational losses to be small, but recent infrared camera technology has shed new light on these sources of vented gas, particularly from condensate storage tanks.[16] According to oil and gas industry representatives, the cameras helped reveal that losses from storage tanks and fugitive emissions were much higher than they originally thought (video link: http://www.gao.gov/products/GAO-11-34).[17] In addition, recent calculations from EPA suggest that emissions from completions and liquid unloading make larger contributions to lost gas than previously thought possible. Operators can use a number of techniques to estimate emissions based on gas and oil characteristics and well operating conditions, such as temperature and pressure, without taking direct measurements of escaping gas.

While venting and flaring of natural gas is often a necessary part of production, the lost gas has both economic and environmental implications. On federal oil and gas leases, natural gas that is vented or flared during production, instead of captured for sale, represents a loss of royalty revenue for the federal government.[18] Venting and flaring natural gas also adds to greenhouse gases in the atmosphere. In general, flaring emits CO_2, while venting releases methane, both of which the scientific community agrees are contributing to global warming. Methane is considered particularly harmful in this respect, as it is roughly 25 times more potent by weight than CO_2 in its ability to warm the atmosphere over a 100-year period and almost 72 times more potent over a 20-year period, according to the Intergovernmental Panel on Climate Change.[19] Other hydrocarbons and compounds in vented and flared gas can also harm air quality by increasing ground-level ozone levels and contributing to regional haze. Volatile organic compounds, present in vented gas, are contributors to elevated ozone and haze, and ozone is a known carcinogen, according to EPA analysis.[20] In some areas in the western United States, the oil and gas industry is a major source of volatile organic compounds. According to EPA, in many western states, including in many rural areas where there is substantial oil and gas production and limited population, there have been increases in ozone levels, often exceeding federal air quality limits.[21] Interior is required to conduct environmental impact assessments in advance of oil and gas leasing and generally works with state environmental and air quality agencies to ensure that oil and gas producers will comply with environmental laws such as the Clean Air Act or Clean Water Act and the related implementing regulations. However, the state agencies may be charged with maintaining the standards established by the federal government in law and regulation, and often have primary responsibility in this regard.[22]

While much of the natural gas that is vented and flared is considered to be unavoidably lost, certain technologies and practices can be applied throughout the production process to capture some of this gas according to the oil and gas industry and EPA. The technologies' technical and economic feasibility varies and sometimes depends on the characteristics of the production site. For example, some technologies require a substantial amount of electricity, which may be less feasible for remote production sites that are not on the electrical grid. However, certain technologies are generally considered technically and economically feasible at particular production stages, including the following:[23]

- *Drilling:* Using "reduced emission" completion equipment when cleaning out a well before production, which separates mud and debris to capture gas or condensate that might otherwise be vented or flared.
- *Production:* Installing a plunger lift system to facilitate liquid unloading. Plunger-lift systems drop a plunger to the bottom of the well, and when the built-up gas pressure pushes the plunger to the surface, liquids come with it. Most of the accompanying gas goes into the gas line rather than being vented. Computerized timers adjust when the plunger is dropped according to the rate at which liquid collects in the well, further decreasing venting.
- *Storage:* Installing vapor recovery units that capture gas vapor from oil or condensate storage tanks and send it into the pipeline.
- *Dehydration:* Optimizing the circulation rate of the glycol and adding a flash tank separator that reduces the amount of gas that is vented into the atmosphere.[24]
- *Pneumatic devices:* Replacing pneumatic devices at all stages of production that release, or "bleed," gas at a high rate (high-bleed pneumatics) with devices that bleed gas at a lower rate (low-bleed pneumatics).

In 2004, we reported that information on the extent to which venting and flaring occurs was limited.[25] Although BLM and BOEMRE require operators to report data on venting and flaring on a monthly basis, our 2004 report found that these data did not distinguish between gas that is vented and gas that is flared, making it difficult to accurately identify the extent to which each occurs. In implementing our recommendations for offshore operators, BOEMRE now requires operators to report venting and flaring separately and to install meters to measure this gas on larger platforms.[26] The Energy Information Administration (EIA)[27] also collects data from oil and gas producing states on venting and flaring, but our 2004 work found that EIA did not consider these state-reported data to be consistent and, according to discussions with EIA officials, these data have not improved.

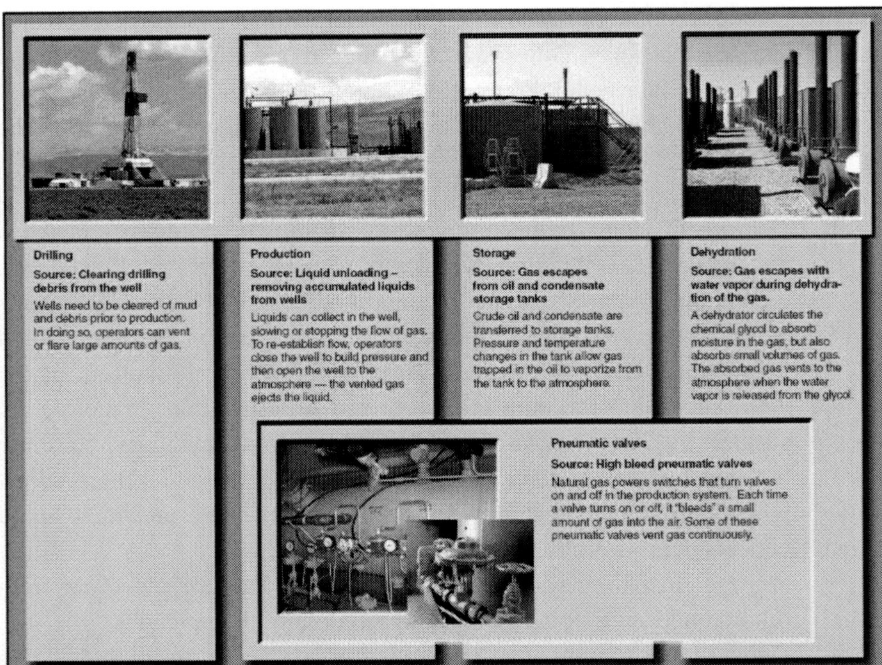

Source: GAO; photos from Bureau of Land Management; Storage photo from Wyoming Department of Environment Quality.

Figure 1. Illustrative Example of Onshore Production and Associated Sources of Vented and Flared Gas.

AVAILABLE ESTIMATES OF VENTED AND FLARED NATURAL GAS VARY, BUT VOLUMES REPORTED TO OGOR ARE LIKELY UNDERESTIMATED BECAUSE THEY DO NOT INCLUDE SOME SOURCES

Available estimates of vented and flared natural gas on federal leases vary considerably, and we found that estimates based on data from MRM's OGOR data system likely underestimate these volumes because they include fewer sources of emissions than other estimates, including EPA's and WRAP's. For onshore federal leases, operators reported to OGOR that about 0.13 percent of the natural gas produced was vented and flared, while EPA estimates showed the volume to be about 4.2 percent, and estimates based on WRAP data showed it to be as high as 5 percent. Similarly, for offshore federal leases, operators reported to OGOR that 0.5 percent of the natural gas produced was vented and flared, while data in BOEMRE's GOADS system—a database that focuses on the impacts of offshore oil and gas exploration, development, and production on air quality in the Gulf of Mexico region—showed that volume to be about 1.4 percent, and estimates from EPA showed it to be about 2.3 percent.[28]

Onshore Leases

Onshore leases showed the largest variation between OGOR data and others' estimates of natural gas venting and flaring. Operators reported to MRM's OGOR system that about 0.13 percent of the natural gas produced on onshore federal leases was vented or flared each year between 2006 and 2008.[29] BLM uses guidance from 1980, which sets limits on the amount of natural gas that may be vented and flared on onshore leases, requires operators to report vented and flared gas to OGOR, and in some cases to seek permission before releasing gas.[30] Although the guidance states that onshore operators must report all volumes of lost gas to OGOR, it does not enumerate the sources that should be reported or specify how they should be estimated.[31] Staff from BLM told us that the reported volumes were from intermittent events like completions, liquid unloading, or necessary releases after equipment failures; however, operators did not report operational sources such as venting from oil storage tanks, pneumatic valves, or glycol dehydrators. In general, BLM staff said that they thought that vented and flared gas did not represent a significant loss of gas on federal leases. In addition, we found a lack of consistency across BLM field offices regarding their understanding of which intermittent volumes of lost gas should to be reported to OGOR. For example, staff from some of the offices said that they thought that intermittent vented and flared gas was not to be reported if operators had advance permission or where volumes were under BLM's permissible limits, while others said that they thought that operators still needed to report this gas. Our discussions with operators reflected this lack of consistency from BLM field office staff. Operators we spoke with said that they generally did not report operational sources, and in some cases did not report intermittent sources as long as they were under BLM's permissible limits for venting and flaring.[32]

In contrast, EPA's estimate of venting and flaring was approximately 4.2 percent of gas production on onshore federal leases for the same period and consistently included both intermittent and operational sources. EPA estimated these emissions using data on average nationwide oil and gas production equipment and their associated emissions (see table 1).[33] As noted earlier, venting from operational sources had not previously been seen as a significant contributor to lost gas. With these additional sources, EPA's estimates are around 30 times higher than the volumes operators reported to OGOR. According to EPA's estimates, the amount of natural gas vented and flared on onshore leases totaled around 126 billion cubic feet (Bcf) of gas in 2008. This amount is roughly equivalent to the natural gas needed to heat about 1.7 million homes during a year, according to our calculations. See figure 2 for a comparison between EPA's estimated gas emissions and the volumes reported to OGOR as a percentage of gas production on federal onshore leases.

Similarly, analysis of WRAP data for five production basins in the mountain west in 2006 indicated as much as 5 percent of the total natural gas produced on federal leases was vented and flared. WRAP based its estimates, in part, on a survey of the types of equipment operators were using,[34] and provided a detailed list of sources to be reported. WRAP's data included similar sources as EPA's data, as well as estimates of emissions from fugitive sources like leaking seals and valves. Although estimates based on WRAP data varied from basin to basin—between 0.3 and 5 percent—they were consistently much higher than the volumes operators reported to OGOR. The average vented and flared gas as a percentage of production was 2.2 percent across the five basins.[35] See table 2 for a list of the key sources in one of the five basins.

Table 1. EPA's Estimates of Vented and Flared Natural Gas and Sources for Onshore Federal Leases

Sources (2008)	Volume (Bcf)
Flared (variety of sources)	28
Pneumatic devices	16
Gas well liquid unloading	17
Well completions	30
Oil and condensate storage tanks	18
Glycol dehydrators	7
Other	10
Total	126

Source: GAO analysis of EPA data.

Note: Volatile organic compounds accounted for 26 Bcf of these emissions and were mostly from storage tanks and dehydrators according to EPA.

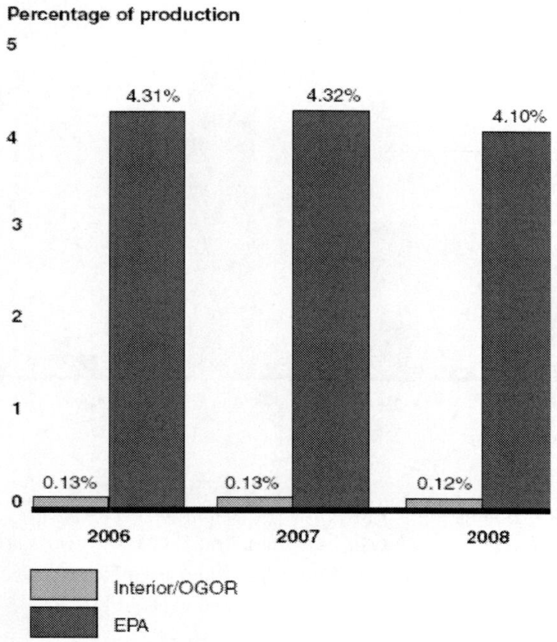

Source: GAO analysis of EPA and OGOR data for federal leases.

Figure 2. Comparison of OGOR Reported Volumes to EPA's Estimates of Vented and Flared Natural Gas for Onshore Federal Leases.

Table 2. Estimates of Vented and Flared Natural Gas based on 2006 WRAP Data for Federal Leases in the Piceance Basin (Colo.)

Sources from Piceance Basin	Volume (Bcf)
Well completions	2.4
Pneumatic devices	0.5
Gas well liquid unloading	0.4
Fugitive emissions	0.1
Condensate storage tanks	0.1
Other sources	0.4
Total	3.8

Source: Environ Corp. analysis of 2006 WRAP data for federal leases.

Note: Flared gas is included throughout several source categories, including completions and storage tanks. We chose to present data from the Piceance basin because it was representative of the key sources common to the other basins. Volume figures in table do not sum to 3.8 Bcf due to rounding.

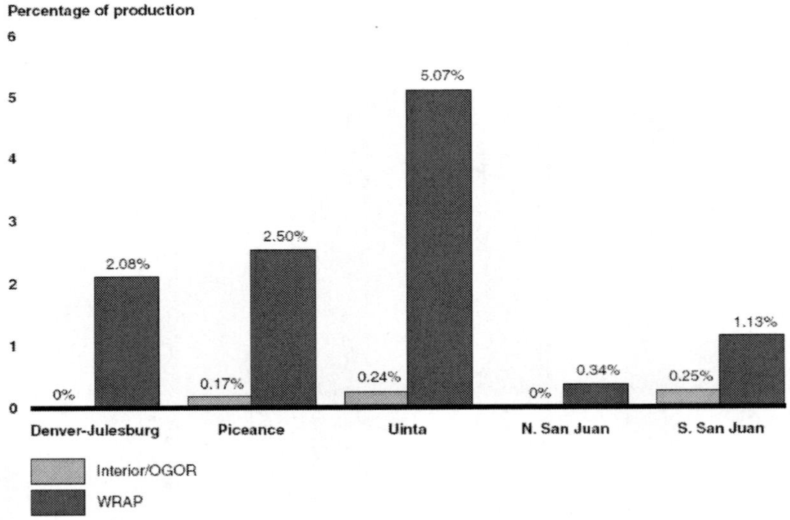

Source: GAO analysis of 2006 WRAP and OGOR data for federal

Note: The Denver-Julesburg, North San Juan, and Piceance basins are in Colorado, the Uinta basin is in Utah, and the South San Juan is in New Mexico. See appendix I for a map of the basins.

Figure 3. Comparison of 2006 OGOR-Reported Volumes to Estimates Based on 2006 WRAP Data of Vented and Flared Natural Gas for Onshore Federal Leases in Five Basins.

In figure 3, which compares estimates based on WRAP data with the volumes operators reported to OGOR for 2006, for the Uinta basin, the WRAP estimate was about 20 times higher than the volumes reported to OGOR, and for two other basins (i.e., Denver-Julesburg and N. San Juan) no volumes of vented and flared gas were reported to OGOR.[36]

Offshore Leases

Offshore leases showed less variation between OGOR data and others' estimates of natural gas venting and flaring than onshore leases, but the volumes that operators reported to MRM's OGOR were still much lower than the volumes they reported to BOEMRE's GOADS system and estimates from EPA. Operators reported to OGOR that between 0.3 and 0.5 percent of the natural gas produced on offshore leases was vented and flared each year from 2006 to 2008; however, they reported to GOADS that they vented and flared about 1.4 percent—about 32 Bcf—of the natural gas produced on federal leases in the Gulf of Mexico in 2008.[37]

Although regulations require offshore operators to report all sources of lost gas to OGOR, BOEMRE officials said that that this did not include fugitive emissions. Furthermore, these officials also said that operators likely reported volumes from some operational sources as "lease-use" gas instead of including it in the venting and flaring data, thus contributing to the differences between OGOR and GOADS.[38] GOADS data included sources similar to those included in EPA's and WRAP's data for onshore production, including the same operational sources. Further, guidance to operators for reporting to GOADS explicitly outlines the sources to be reported and how they should be estimated, while guidance for OGOR does not.

Table 3 outlines the emission sources for volumes operators reported to the GOADS system for 2008. In addition, EPA's offshore estimates showed that around 2.3 percent of gas produced on offshore federal leases—as much as 50 Bcf—was vented and flared every year from 2006 to 2008. According to our analysis of EPA's work, additional venting from natural gas compressors, used to maintain proper pressure in production equipment, accounted for the majority of the difference between the offshore EPA and GOADS volumes.[39]

Table 3. GOADS's Reported Vented and Flared Natural Gas and Sources for Offshore Federal Leases

Sources from 2008 GOADS Study	Volume (Bcf)
Venting	12
Flaring	7
Fugitive emissions	6
Pneumatic devices	3
Glycol dehydrators	1
Other sources	2
Total	32

Source: GAO analysis of GOADS data.

Note: Volume figures in table do not sum to 32 Bcf due to rounding. See appendix I for more information on these figures.

On several occasions BOEMRE has made comparisons between data on vented and flared volumes in the OGOR and GOADS systems, according to BOEMRE officials. In 2004, BOEMRE compared data from the 2000 GOADS study with data from OGOR for a subset of offshore leases and found reported vented and flared volumes were not always in agreement— attributing this difference to different operator interpretations of GOADS and OGOR reporting requirements. BOEMRE officials said they revised reporting procedures for the 2005 GOADS study. More recently, BOEMRE made similar comparisons between data from the 2008 GOADS study and OGOR data for a subset of leases and found they were in closer agreement.[40] BOEMRE officials told us they will continue to make such comparisons to try to ensure the accuracy of the data in each system. In reporting volumes of vented and flared gas to both systems, operators can choose from a broad array of software packages, models, and equations to estimate emissions, and these techniques can yield widely varied results. For example, one study found that various estimation techniques to determine emissions from oil storage tanks either consistently underestimated or overestimated vented volumes.[41] OGOR reporting instructions for both onshore and offshore operators, as noted, do not specify how operators should estimate these volumes.

As part of our review, we analyzed 2008 OGOR and GOADS data for the Gulf of Mexico and found that the OGOR data likely underestimated the volumes of vented and flared natural gas on federal offshore leases. To do this analysis, we compared 2008 data from GOADS's vent and flare source categories with OGOR data for the same categories—looking at these source categories allowed us to directly compare the two data systems. In doing this analysis, we accounted for OGOR's exclusion of fugitive emissions and the reporting of sources, like pneumatic valves, as lease-use gas. Our analysis found that the volumes operators reported to OGOR— about 12 Bcf—were much lower than the volumes operators reported to GOADS—about 18 Bcf. Neither we nor MRM and BOEMRE officials could account for or explain these differences in the two data systems. BOEMRE officials said that they are still working to improve reporting to OGOR and GOADS and expect these two data systems to converge in the future.

To improve reported data, BOEMRE recently released a final rule, in response to the recommendations in our 2004 report, that requires operators on larger offshore platforms to route vented and flared gas from a variety of sources through a meter to allow for more accurate measurement, among other things.[42] BOEMRE officials said that these meters would help to improve the accuracy of data reported to both OGOR and GOADS.[43] However, BOEMRE officials said they have had to address questions from some operators who were not sure which sources of vented gas should be routed through the newly required meters. In this regard, these officials said it may be useful to enumerate the required emission sources for reporting to OGOR in future guidance to offshore operators. They also noted that BOEMRE is planning a workshop in October 2011 to stress to operators the need for accurate reporting on their submissions to both GOADS and OGOR systems. In a similar way, EPA has taken action to improve the reporting of emissions from the oil and natural gas industry. EPA recently proposed a greenhouse gas reporting rule that would require oil and gas producers emitting over 25,000 metric tons of carbon dioxide equivalent to submit detailed data on vented and flared gas volumes to allow EPA to better understand the contribution of venting and flaring to national greenhouse gas emissions. For onshore leases, the proposed EPA rule provides details on the specific sources of vented and flared gas to be measured and proposes standardized methods for estimating volumes of greenhouse gas emissions where direct measurements are not possible. For offshore leases, operators would use the GOADS system to report venting and flaring. Data collection would begin in 2011 if the rule becomes finalized in 2010.

AVAILABLE TECHNOLOGIES COULD REDUCE ABOUT 40 PERCENT OF NATURAL GAS ESTIMATED TO BE LOST TO VENTING AND FLARING ON ONSHORE FEDERAL LEASES, POTENTIALLY INCREASING FEDERAL ROYALTY PAYMENTS AND REDUCING GREENHOUSE GAS EMISSIONS

Data from EPA, supported by information obtained from technology vendors and our analysis of WRAP data, suggest that about 40 percent of natural gas estimated to be vented and flared on federal onshore leases could be economically captured with currently available control technologies, although some barriers to their increased use exist. Such captures could increase federal royalty payments and reduce greenhouse gas emissions.

Available Technologies Could Reduce Venting and Flaring on Onshore Federal Leases, but According to EPA Officials and Technology Vendors, Some Barriers Exist

Available technologies could reduce venting and flaring at many stages of the production process. However, there are some barriers to implementing these technologies.

Available Technologies Could Reduce Venting and Flaring on Onshore Federal Leases

EPA analysis and our analysis of WRAP data identified opportunities for expanded use of technologies to reduce venting and flaring. Specifically:

- EPA's 2008 analysis, the most recent data available, indicates that the increased use of available technologies, including technologies that capture emissions from sources such as well completions, liquid unloading, or venting from pneumatic devices,[44] could have captured about 40 percent—around 50 Bcf—of the natural gas EPA estimated was lost from onshore federal leases nationwide. [45] For instance, EPA found significant opportunities to add "smart" automation to existing plunger lifts, which tune plunger lifts to maximum efficiency and, in turn, minimize the amount of gas lost to venting. EPA estimated that using this technology where economically feasible could have resulted in the capture of more than 7 Bcf of vented and flared natural gas on federal leases in 2008—around 6 percent of the total volume estimated by EPA to be vented and flared on onshore federal leases. Similarly, EPA estimated that additional wells on onshore federal leases could have incorporated reduced emission completion technologies in 2008, which could have captured an additional 14.7 Bcf of vented and flared natural gas. Table 4 outlines EPA's estimates of potential reductions in venting and flaring on onshore federal leases.

Table 4. Potential Nationwide Reductions on Onshore Federal Leases from Increased Use of Venting and Flaring Reduction Technologies, 2008

Emission source	Potential reduction (Bcf)	Percent of totalvolume EPA estimatedvented and flared
Gas well liquid unloading	7.2—expand use of smart automated plungers	5.7
Well completions	14.7—expand use of reduced emission completions	11.7
Glycol dehydrators	5.7—install vapor recovery devices	4.5
Pneumatic devices	9.7—use low-bleed devices	7.7
Oil and condensate storage tanks	12.9—install vapor recovery units	10.2
Total	50.2	39.8

Source: GAO analysis of EPA data.

- Our analysis of WRAP's 2006 data from certain onshore production basins also highlighted the possibility for additional venting and flaring reductions. We found significant regional differences in use of control technologies in oil and gas production basins in the mountain west and, subsequently, differences in vented and flared volumes, as a percentage of total basin production. For example, according to the WRAP data, most pneumatic devices in the Piceance basin in northwest Colorado were low-bleed in 2006, while high-bleed pneumatic devices were still predominant in the neighboring Uinta basin in northeast Utah. Although the Piceance and Uinta basins are part of the same geological formation and share many characteristics, including type of gas development and extraction methods, the WRAP data show that the venting and flaring volumes on federal leases in the Uinta basin are nearly double those in the Piceance basin as a percentage of total gas production. See appendix I for a map of these basins.

Differing rates of use of venting and flaring reduction technologies— among states, oil and gas production basins, and individual operators— can be attributed to two key factors based on our analysis. Variations in state air quality regulations are one factor, according to EPA and state agency officials and industry representatives. For example, Colorado has stricter requirements than Utah for emissions controls, according to officials, which partly explains the variation in levels of control technologies in use in these production basins. Similarly, over the last 5 years Wyoming has instituted many regulatory changes to address increases in ground-level ozone, including stricter emission-reduction requirements for storage tanks and reduced emission completions.

The net economic benefit of installing equipment to capture vented and flared gas is the second key factor. The cost of implementing a given technology can be substantial—and may be especially burdensome for smaller operators. Nonetheless, in many cases, the costs are recovered quickly as the captured gas is sold, according to industry and EPA officials. According to documents from EPA's Natural Gas STAR program, a voluntary partnership with industry to encourage reductions in gas venting, in many cases, the cost of implementing these control technologies can be recovered in less than 1 year. According to EPA and industry representatives, for operators with sufficient resources— including engineering and cost-estimation teams, as well as capital for infrastructure—decisions to potentially install capture equipment are easy to make based on simple economic considerations. For example, the cost of switching from high-bleed to low-bleed pneumatic devices ranges from $700 to $3,000 per device, which can be recovered in 2 to 8 months, on average, according to EPA documents. Similarly, retrofitting an oil production storage tank with a vapor recovery unit can cost tens of thousands of dollars, but the gas saved can pay for the technology generally within 2 years. According to EPA and BLM officials, some operators have already implemented plunger-lift systems, vapor recovery units, reduced emission completions, and other technologies due to the economic benefit of doing so. However, these officials cautioned that the return-on-investment calculation can be complicated by a number of factors, including the geology and location of the production basin and the differences in the composition of extracted oil or gas. For instance, while some high-bleed pneumatic devices vent more gas than low-bleed devices, the higher bleed rates keep the equipment from freezing in cold conditions, according to industry officials. Similarly, reduced emission completions are not economically viable for conventional gas wells with low wellhead pressures, as the costs of reduced emission completion equipment can easily outweigh the benefits of capturing the gas, according to industry representatives. In addition, EPA and industry officials told us that installing these technologies may require other significant infrastructure investments, such as a new pipeline from an oil well where natural gas is currently being vented or flared to a gas sales line, which could make the investment in these technologies cost-prohibitive.

A number of industry representatives and EPA officials noted the potential for currently-developing carbon markets to influence the economics of venting and flaring control technologies. Carbon markets generally refer to real financial markets where carbon emission reductions, known as carbon offset credits, are bought and sold; companies that are emitting more than the amount of carbon allocated, or allowed by the government regulators, can buy credits to offset their excess emissions and companies that reduce emissions can sell those reductions as credits. Operators are increasingly able to document the carbon reductions achieved through installations of the technologies and, in turn, sell these offset credits on open carbon markets according to industry officials. Potential opportunities to claim and sell these carbon offset credits may add to the economic incentives for using these control technologies, according to some industry officials.[46] Although there is some risk involved with claiming these offsets, as the markets are developing, pending federal legislation could make it more likely that a market for carbon will become increasingly relevant, according to industry officials.

The experiences of some operators in implementing technologies to reduce venting and flaring show the economic advantages that can result. For example, BP installed smart-automated plunger lifts on its onshore wells throughout the San Juan Basin and reported achieving a 99 percent reduction in vented volumes as a result, increasing production and profits, according to company representatives. In addition, according to company officials, the company has replaced all of its high-bleed pneumatic devices with low-bleed devices, saving approximately 3.4 Bcf of natural gas emissions annually. BP also reported investing $1.2 million in reduced emission completions since

2000, which it credits with saving 1.5 Bcf of gas and almost 29,000 barrels of condensate. These emissions savings prevented over 100,000 metric tons of CO_2 and 2,000 metric tons of methane from entering the atmosphere and increased revenues by almost $5.8 million, according to company documents. The results of BP's actions were likely a factor in the estimates of venting and flaring based on WRAP data for the North San Juan basin that we reported earlier; BP was the major operator in that basin, and it had the lowest estimates of venting and flaring. Similarly, Devon Energy recently took steps to expand its use of venting and flaring reduction technologies for some of its onshore wells that have resulted in significant successes, according to company representatives. In 2008, Devon representatives reported 10.4 Bcf of methane emission reductions that they attribute to the replacement of high-bleed pneumatic devices, installation of vapor recovery units on storage tanks, use of automated plunger lift systems, and use of reduced emission completions, among other technologies. Overall, the company saved more than $125 million since 1990 by implementing these technologies, according to Devon representatives.

Vendors of these technologies also cited success stories. For example, one vendor cited an example where installation of two vapor recovery units onshore requiring capital investments of more than $20,000 yielded a full return on investment in less than a month through the capture of otherwise vented gas. In addition, one operator replaced and retrofitted 400 high-bleed pneumatics on wells onshore, at a cost of more than $118,000, but found an annual savings in captured gas of nearly $149,000, for a payback on investment in less than 1 year. Each of these cases demonstrates that the venting and flaring reduction technologies can, under the right circumstances, add to an operator's bottom line. These results illustrate the potential multiple benefits of venting and flaring reduction technologies—benefits to industry in the form of additional revenues; benefits to the government in the form of increased royalty payments; and benefits to the environment in the form of reductions in vented and flared greenhouse gases.

EPA Officials and Technology Vendors Identified Barriers to Implementing Available Technologies

Despite the potential economic benefits of using these technologies, there are barriers to their implementation for some operators, according to an EPA official and technology vendors. One key barrier is that many operators are unaware of the economic advantages. In part, this is because smaller operators often do not have the time or expertise to undertake the engineering analysis to understand whether and how they can benefit, according to EPA and technology vendors. Also, these officials said that smaller operators often do not have the capital to purchase equipment, regardless of whether they can recover the costs. According to officials, the voluntary nature of the EPA Natural Gas STAR program is not enough to spur industry to change, and one industry official stated that the sometimes contentious relationship between the federal government and private industry contributes to this lack of awareness. Private industry does not always take federal efforts to encourage industry to alter business practices at face value, according to officials. One industry representative cited reluctance to participate in EPA's Natural Gas STAR program as an example of this skepticism.

A number of other factors can also contribute to operators not adopting venting and flaring reduction technologies. Officials that we spoke with said that overcoming "institutional inertia"—a company's tendency to do business and carry out operations as it always has—is key to adopting these technologies. In a similar vein, industry and EPA officials told us that upper management support is critical for these types of efforts to go forward, and many companies' management is focused on other efforts that are deemed more important than what are seen as incremental improvements in operations. For example, the operator may choose to invest its limited available capital in drilling a new well, which may have a larger return than investments in capturing vented or flared gas from an existing well, according to industry representatives.

Reductions in Natural Gas Lost to Venting and Flaring Could Increase Federal Royalty Payments and Reduce Greenhouse Gas Emissions

Reductions in natural gas lost to venting and flaring from federal leases would increase the volume of natural gas produced and sold, thereby potentially increasing federal royalty payments. If, for instance, a total of 126 Bcf of natural gas was lost to venting and flaring on onshore federal leases in 2008, as EPA has estimated, that loss would equal approximately $58 million in federal royalty payments. If, as EPA estimates, 40 percent of this lost gas could have been economically captured and sold, federal royalty payments could increase by approximately $23 million annually, which represents about 1.8 percent of annual federal royalty payments on natural gas.[47]

Reducing natural gas lost to venting and flaring from federal leases could also reduce greenhouse gases to the atmosphere according to our calculations. Because methane is about 25 times more potent as a greenhouse gas over a 100-year period, and almost 72 times more potent over a 20-year period according to the Intergovernmental Panel on Climate Change,[48] reducing direct venting of natural gas to the atmosphere has a significantly greater positive effect, in terms of global warming potential, than does reducing flaring. Again using EPA's estimates, if a total of 98 Bcf of natural gas was vented and 28 Bcf was flared annually, those releases would account for about 41 million metric tons of carbon dioxide equivalent released to the atmosphere, which would be roughly equivalent to the emissions of almost 8 million passenger vehicles or about 10 average-sized coal-fired power plants. Capturing 40 percent of this volume would result in emissions reductions of about 50 Bcf, which is equivalent to the emissions of 3.1 million passenger vehicles or about 4 average-sized coal-fired power plants, according to our analysis.[49]

Some EPA officials also told us that they believed that federal efforts to reduce venting and flaring could also have a spillover effect—that is, it could lead operators to use these technologies on state and private leases as well. Data from EPA and WRAP included vented and flared gas from nonfederal leases, and the data showed that there were similar percentages of gas being lost, suggesting that the potential greenhouse gas reductions from the expanded use of these technologies could go well beyond those from federal oil and gas production.[50]

We did not find complete quantitative data on reduction opportunities offshore from Interior, EPA, or others that could be used to fully identify the potential to reduce emissions offshore. However, EPA officials told us that opportunities for reducing emissions from venting and flaring from offshore production platforms likely exist. For instance, EPA found that various production components, including valves and compressor seals, contribute significant volumes of fugitive emissions, but that these emissions could be mitigated through equipment repair or retrofitting. One estimate based on EPA analysis of 15 offshore platforms in 2008, suggests that most of the gas lost through compressor seals could be recovered economically—saving about 70 percent of the overall gas they estimated to be lost on those platforms. However, EPA's analysis warns that some mitigation strategies may be less cost-effective in the offshore environment because capital costs and installation costs tend to be higher.

Interior's Oversight Does Not Ensure that Operators Minimize Venting and Flaring on Federal Leases, While a Voluntary EPA Program Has Reduced Vented Gas According to EPA and Industry Participants

Interior is responsible for ensuring that operators minimize natural gas venting and flaring on federal onshore and offshore leases; however, while both BLM and BOEMRE have taken steps to minimize venting and flaring on federal leases, their oversight of such leases has several limitations. Although EPA does not have a direct regulatory role with respect to managing federal oil and gas leases, its Natural Gas STAR program has helped to reduce vented gas on federal leases according to EPA and industry participants.

BLM and BOEMRE Have Taken Steps to Minimize Venting and Flaring on Federal Leases, but Their Oversight Has Several Limitations

As part of their oversight responsibilities, Interior's BLM and BOEMRE are charged with minimizing the waste of federal resources, and, to that end, both agencies have issued regulations and guidance that limit venting and flaring of gas during routine procedures such as liquid unloading and well completions.[51] However, their oversight has several limitations, namely (1) the regulations and guidance do not address new capture technologies or all sources of lost gas; (2) the agencies do not assess options for reducing venting and flaring in advance of oil and gas production for purposes other than addressing air quality; and (3) the agencies have not developed or do not use information regarding available technologies that could reduce venting and flaring.

Regulations and Guidance Limit Venting and Flaring, but Do Not Address Newer Technologies or All Sources of Lost Gas

Onshore Leases

BLM's guidance limits venting and flaring from routine procedures and requires operators to request permission to vent and flare gas above these limits.[52] If operators request permission to exceed these limits, BLM is to assess the economic and technical viability of capturing additional gas and require its capture when warranted.[53] Although BLM guidance sets limits on venting and flaring of natural gas and allows flexibility to exceed them in certain cases, it does not address newer technologies or all sources of lost gas. Specifically, BLM guidance is 30 years old and therefore does not address venting and flaring reduction technologies that have advanced since it was issued. For example, since the guidance was written, technologies have been developed to economically reduce emissions from well completions and liquid unloading—namely the use of reduced emission completion and automated plunger lift technologies respectively. These two sources of emissions were important contributors to vented and flared volumes that we discussed earlier. Despite this fact, the use of such technologies where it is economic to do so is not covered in BLM's current guidance. In general, BLM officials said that they thought the industry would use venting and flaring reduction technologies if they made economic sense. Similarly, new lower-emission devices could also reduce venting and flaring from other sources of emissions that are not covered by BLM's guidance, such as pneumatic valves or gas dehydrators—two sources that contribute to significant lost gas. In discussions with BLM staff about their guidance, staff acknowledged that existing guidance was outdated given current technologies and said that they were planning to update it by the second quarter of 2012.

Offshore Leases

Like BLM, BOEMRE has regulations that limit the allowable volumes of vented and flared gas from offshore leases to minimize losses of gas from routine operations. Operators can also apply for permission to exceed these limits and, like BLM, BOEMRE would evaluate the economic and technical viability of capturing additional gas. Further, BOEMRE inspects offshore platform facilities each year and, as part of these inspections, reviews on-site daily natural gas venting records.[54] BOEMRE officials told us that the agency requires operators to keep these venting records and that it uses them to, among other things, identify any economically viable opportunities for an operator to install control equipment. Overall BOEMRE officials said that operators were required to install venting and flaring reduction equipment where economic, even if they would make as little as $1 in net profit from the captured gas. According to agency officials, due to the type of production and operations offshore, reduction opportunities mostly consist of installing vapor recovery units, and these officials said that they generally believe that companies have installed such equipment where it is economic to do so.

Although BOEMRE conducts regular inspections, the daily venting records do not include all sources of vented gas. For example, emission estimates from sources of gas such as pneumatic valves and glycol dehydrators are not included, and therefore inspectors are not able to make assessments of the potential to reduce emissions from these sources. Both of these sources were contributors to lost gas offshore from the 2008 GOADS study, suggesting potential reduction opportunities. BOEMRE officials said that the agency considers these sources lease-use gas, and as a result, believed that they could not legally consider the economic and technical viability of this gas and require its capture when warranted. However, based on our review of BOEMRE regulations and authorizing legislation, it appears that BOEMRE has the authority to require operators to minimize the loss of this gas, including requiring its capture where appropriate. BOEMRE officials agreed with our assessment.

The Agencies Do Not Assess Options for Reducing Venting and Flaring in Advance of Oil and Gas Production

Onshore Leases

While BLM regulations authorize and direct BLM officials to offer technical advice and issue orders for specific lease operations to minimize waste,[55] BLM does not explicitly assess options to minimize waste from vented and flared gas before production. For example, we identified two phases in advance of production where BLM could assess venting and flaring reduction options—during the environmental review phase and when the operator applies to drill a new well. However, the agency does not explicitly assess these options, or discuss them with operators, during either phase. For example, during the environmental review phase, BLM works with states to assess emissions from oil and gas production, and that air quality assessment may include venting and flaring reduction requirements. According to BLM officials, since states generally have primary responsibility to implement and enforce air quality standards, the standards drive these requirements, and states focus only on the role venting and flaring plays in air pollution, rather than the minimization of waste. Therefore in production basins where air quality standards are being met, or where only minimal use of technology is required to meet them, BLM would not assess venting and flaring reduction technologies to the full extent that they could economically reduce vented and flared gas.[56] One official noted that some BLM officials felt constrained in their ability to consider the use of venting and flaring reduction technologies because of this. Similarly, during the phase when operators apply to drill new wells, BLM assesses detailed technical and environmental aspects of the project, but BLM officials told us their assessment does not include a review of options to reduce venting and flaring.[57]

Offshore Leases

Similar to BLM, BOEMRE assesses venting and flaring reduction options in advance of production to determine whether vented and flared gas from offshore platforms would harm coastal air quality, but again, the focus is on meeting air quality standards rather than assessing whether gas can be economically captured. Therefore, when BOEMRE does not anticipate harm to coastal air quality, as is often the case according to officials, the agency does not further consider venting and flaring reduction options at this phase. Further, while the application operators submit in advance of drilling must include a description of the technologies and recovery practices that the operator will use during production,[58] venting and flaring reduction options are not included in that submission.

Agencies Have Not Developed or Do Not Use Information Regarding Available Technologies that Could Reduce Venting and Flaring

Onshore Leases

We found that BLM does not maintain a database regarding the extent to which available venting and flaring reduction technologies are used on federal oil and gas leases.[59] As such, it could be difficult for BLM to identify opportunities to reduce venting and flaring or estimate the potential to increase the capture of gas that is currently vented or flared. For example, while BLM guidance provides that the natural gas vaporizing from storage tanks must be captured if BLM determines recovery is warranted, BLM does not collect data on the use of control technologies and available OGOR data do not contain the volumes of lost gas from storage tanks. Thus BLM may be overlooking circumstances where recovery could be warranted. In addition, according to BLM officials we spoke with, although infrared cameras can be used to identify sources of lost gas, BLM has not used them during inspections of production facilities. Although relatively expensive, infrared cameras allow users to rapidly scan and detect vented gas or leaks across wide production areas. BLM officials cited budgetary constraints and challenges in developing a policy and protocols for why the cameras have not been used regularly by the agency.

Offshore Leases

Although the GOADS data system contains some information on the types of equipment operators use, BOEMRE has not analyzed this information to identify emission-reduction opportunities according to officials. GOADS contains information about the use of equipment such as vapor recovery systems. These data have not been used by BOEMRE to identify venting and flaring reduction opportunities because the agency has not considered using these data for purposes other than addressing air quality, according to a BOEMRE official. Nonetheless, based on our review of the GOADS data system, by not analyzing such data, BOEMRE is not able to identify emission-reduction opportunities. As a case in point, we found that emissions from pneumatic valves in the 2008 GOADS study made noticeable contributions to overall lost gas, which might suggest the potential to expand the use of low-bleed pneumatics in some cases. BOEMRE officials also noted that, unlike BLM, its inspectors had used infrared cameras to look for obvious sources of vented and flared gas in a few sample locations close to shore. In this regard, they said expanded use of infrared cameras could be useful to help enforce their new rule that requires the use of meters for vented and flared gas. Specifically, they said that the cameras could identify sources of gas that operators may have not routed through the meter as required. They also noted that expanded use of the cameras could help to identify and potentially reduce fugitive gas emissions that currently go undetected.[60]

EPA's Voluntary Natural Gas STAR Program Has Helped Reduce Vented Gas, According to EPA and Industry Participants

Although Interior has the primary role in federal oil and gas leasing, EPA's Natural Gas STAR program has encouraged some operators to adopt technologies and practices that have helped to reduce methane emissions from the venting of natural gas, according to EPA and industry participants. Through this program, industry partners evaluate their emissions and consider ways to reduce them, although the reductions are voluntary. The program also maintains an online library of technologies and practices to reduce emissions that quantify the costs and benefits of each emission-reduction option. Natural Gas STAR also sponsors conferences to facilitate information exchange between operators regarding emissions reductions technologies. Partner companies report annually about their efforts to reduce emissions along with the volumes of the emission reductions.[61]

According to the Natural Gas STAR Web site, domestic oil and gas industry partners reported more than 114 Bcf of methane emission reductions in 2008, which amounts to about 0.4 percent of the total natural gas produced that year. However, one industry representative said that, while large and midsize operators were aware of the Natural Gas STAR program, smaller operators were not aware and, even if some smaller operators were aware of the program, they may not have the environmental staff to implement the technologies and practices. Despite the potential usefulness of information from the Natural Gas STAR program to oil and gas producers on federal leases, some of the BLM officials that we spoke with were unfamiliar with Natural Gas STAR.

CONCLUSION

Fulfilling its responsibility to ensure that the country's oil and natural gas assets are developed reasonably and result in fair compensation for the American people requires Interior to have accurate and complete information on all aspects of oil and natural gas leases. Interior has collected some information on vented and flared gas through MRM's OGOR system, but without a full understanding of these losses Interior cannot fully account for the disposition of taxpayer resources or identify opportunities to prevent undue waste. MRM's OGOR data system does not provide information on all sources of lost gas, which is the primary source of data that BLM uses to measure overall vented and flared gas onshore. Therefore, OGOR data present an incomplete picture of venting and flaring onshore, leading BLM officials to believe that vented and flared gas volumes do not represent a significant loss of gas on federal leases. Similarly, data in BOEMRE's GOADS data system differ considerably from data in OGOR, and have not been reconciled—raising questions about the accuracy of offshore data sources.

Regarding Interior's oversight of operators venting and flaring gas, because current guidance and regulations from BLM and BOEMRE do not require the minimization of all sources of vented and flared gas—although legislation exists authorizing them to require that waste on federal leases be minimized—operators may be venting and flaring more gas than should otherwise be allowed. In fact, we found that operators are not using available technologies in all cases to economically reduce vented and flared gas. BLM guidance has not kept pace with the development of economically viable capture technologies for a number of sources of lost gas, and BOEMRE has been reluctant to consider the economic and technical viability of minimizing the waste of "lease-use" gas because officials had believed they were legally constrained from doing so.

In addition to the limitations of these regulations, BLM and BOEMRE have not used their authority in two situations where they could potentially further reduce venting and flaring. First, neither agency has used its authority to minimize waste beyond relevant air quality standards by assessing the use of venting and flaring reduction technologies before production. Second, because BLM lacks data about the use of venting and flaring technologies for onshore leases and BOEMRE does not analyze its existing information for offshore leases in its GOADS data system, these agencies are not fully aware of potential opportunities to use available technologies. Further, neither agency takes full advantage of newer infrared camera technology that can help to identify sources of lost gas— as BOEMRE officials have acknowledged, this technology could help reveal additional sources of lost gas.

Ultimately, a sharper focus by BOEMRE and BLM on the nature and extent of venting and flaring on federal leases could have multiple benefits. Specifically, increased implementation of available venting and flaring reduction technologies, to the extent possible, could increase sales volumes and revenues for operators, increase royalty payments to the federal government, and decrease emissions of greenhouse gases. In addition, our analysis of WRAP and EPA data showed as much or more vented and flared gas on nonfederal leases, and we share the observation with EPA officials that a spillover effect may occur, whereby oil and gas producers, seeing successes on their federal leases, take similar steps on state and private leases.

RECOMMENDATIONS FOR EXECUTIVE ACTION

To ensure that Interior has a complete picture of venting and flaring on federal leases and takes steps to reduce this lost gas where economic to do so, we are making five recommendations to the Secretary of the Interior.

To ensure that Interior's data are complete and accurate, we recommend that the Secretary of the Interior direct BLM and BOEMRE to take the following action:

- Take additional steps to ensure that each agency has a complete and accurate picture of vented and flared gas, for both onshore and offshore leases, by (1) BLM developing more complete data on lost gas by taking into consideration additional large onshore sources and ways to estimate them not currently addressed in regulations—sources that EPA's newly proposed greenhouse gas reporting rule addresses—and (2) BOEMRE reconciling differences in reported offshore venting and flaring volumes in OGOR and GOADS data systems and making adjustments to ensure the accuracy of these systems. To help reduce venting and flaring of gas by addressing limitations in their regulations, we recommend that the Secretary of the Interior direct BLM and BOEMRE to take the following four actions:
- BLM should revise its guidance to operators to make it clear that technologies should be used where they can economically capture sources of vented and flared gas, including gas from liquid unloading, well completions, pneumatic valves and glycol dehydrators. BOEMRE should consider extending its requirement that gas be captured where economical to "lease-use" sources of gas;
- BLM and BOEMRE should assess the potential use of venting and flaring reduction technologies to minimize the waste of natural gas in advance of production where applicable, and not solely for purposes of air quality;
- BLM and BOEMRE should consider the expanded use of infrared cameras, where economical, to improve reporting of emission sources and to identify opportunities to minimize lost gas; and
- BLM should collect information on the extent that larger operators use venting and flaring reduction technology and periodically review this information to identify potential opportunities for oil and gas operators to reduce their emissions, and BOEMRE should use existing information in its GOADS data system for this same purpose, to the extent possible.

AGENCY COMMENTS AND OUR EVALUATION

We provided a copy of our draft report to Interior and EPA for review and comment. Interior provided written comments that concurred with four of the five recommendations and partly concurred with the remaining recommendation. Its comments are reproduced in appendix II and key areas are discussed below. EPA did not provide formal comments on the report, but the agency's Office of Air and Radiation provided written comments to GAO staff, which we summarize and discuss below. Interior and EPA also provided other clarifying or technical comments, which we incorporated as appropriate.

Interior's comments reflected the views of BLM and BOEMRE. BLM concurred with all five recommendations and noted that it plans to incorporate recommended actions into its new Onshore Order in order to improve the completeness and accuracy of its data and help address limitations in its current regulations.

BOEMRE concurred with four of the recommendations and partly concurred with our second recommendation that they consider enforcing the economical capture of "lease-use" gas. It stated that we misapprehended the scope of the regulations governing "lease-use" sources of gas in that BOEMRE does not have current regulations to require the capture of "lease-use" gas. In response to this comment, we reworded our recommendation to clarify that BOEMRE should consider extending its existing requirements for the economical capture of gas to "lease-use" gas. In a related point, BOEMRE also noted that we were unable to quantify the potential volumes of additional gas that could be captured by holding operators to this same economic standard for "lease-use" gas. While current data have limitations, BOEMRE's GOADS data suggest potential opportunities to capture additional gas from lease-use sources, namely glycol dehydrators and pneumatic devices. As such, we support BOEMRE's efforts to further evaluate this issue and take action through new guidance or regulations, as it believes appropriate.

EPA's Office of Air and Radiation commented on three areas of the report:

- First, EPA emphasized the significant air quality impacts from the volatile organic compounds (VOC) associated with vented gas and provided us with estimates of the potential volumes of these emissions. While we recognize that the impacts of VOC emissions on air quality are important, these impacts were largely beyond the scope of our work. Nonetheless, we incorporated an estimate of these VOC emissions into supporting notes to table 1 that reflected EPA's estimates of vented and flared gas. We also added additional information to the background regarding VOC emissions.
- Second, EPA suggested that we recommend to BLM and BOEMRE that they require the use of the best available venting and flaring control measures during leasing or drilling permitting. We continue to

believe that BLM and BOEMRE should require the use of these technologies where economical, and recognize that requiring the use of such controls when the economics of capturing gas are unfavorable is not required by current EPA greenhouse gas regulations.

- Third, EPA provided us with its revised emission estimates for vented and flared gas based on updated analysis for its proposed rule on the reporting of greenhouse gases by industry. It also provided us with revised estimates for the use of additional control technologies to reduce the emissions of vented and flared gas. In both cases, we incorporated these revised estimates in our report where applicable.

Frank Rusco
Director,
Natural Resources and Environment

APPENDIX I: OBJECTIVES, SCOPE, AND METHODOLOGY

Our objectives were to (1) examine available estimates of vented and flared natural gas on federal leases; (2) estimate the potential to capture additional vented and flared natural gas with available technologies and the associated potential increases in royalty payments and reductions in greenhouse gas emissions and; (3) assess the federal role in reducing venting and flaring of natural gas.

To examine available estimates of vented and flared natural gas on federal leases, we collected data from the Department of the Interior's (Interior) Bureau of Land Management (BLM), Bureau of Ocean Energy Management, Regulation and Enforcement (BOEMRE), including BOEMRE's Minerals Revenue Management (MRM) program; the Environmental Protection Agency (EPA); and the Western Regional Air Partnership (WRAP). We also interviewed staff from these agencies and oil and gas producers operating on federal leases regarding venting and flaring data collection, analysis, and reporting.

We obtained data from four key sources: MRM's Oil and Gas Operations Report (OGOR) database, BOEMRE's Gulfwide Offshore Activity Data System (GOADS), EPA's Natural Gas STAR Program, and WRAP's analysis of air emissions for a number of western states.

We assessed the quality of the data from each of these sources and determined that these data were sufficiently reliable for the purposes of our report.

MRM provided OGOR data on vented and flared volumes and production for both onshore and offshore federal leases for calendar years 2006 to 2008. MRM uses the OGOR data, in part, to ensure accurate federal royalty payments.[1] The OGOR data are operator-reported, and reported venting and flaring volumes are a mix of empirical measurements and estimates from operators.

MRM was unable to provide complete estimates of vented and flared gas on all federal leases because a portion of federal leases are managed as part of lease agreements—collections of leases that draw from the same oil or gas reservoir, which may include federal and nonfederal leases. MRM was unable to determine the share of reported vented and flared gas from the federal portion of those lease agreements; it reported venting and flaring from (1) lease agreements that included only federal leases and (2) all lease agreements, which included some nonfederal leases. In this report, we discuss the vented and flared volumes from the agreements that contain only federal leases. As a result, we report vented and flared gas volumes from the OGOR data as a percentage of total production on these leases, rather than as absolute volumes, in order to compare the OGOR estimates to estimates from other data sources.

A second source of venting and flaring data was BOEMRE's 2008 GOADS data, which contained estimates of gas lost to venting and flaring on federal leases in the Gulf of Mexico—which accounted for 98 percent of federal offshore gas production in 2008. BOEMRE collects GOADS data every 3 years and uses these data to estimate the impacts of offshore oil and gas exploration, development, and production on onshore air quality in the Gulf of Mexico region. BOEMRE also uses GOADS as part of an impact analysis required by the National Environmental Policy Act. GOADS data capture specific information on a variety of sources of air pollutants and greenhouse gases resulting from offshore oil production. BOEMRE provided us with actual volumes of natural gas released from the vented and flared source categories.

For the other sources, we used the emissions that were reported in GOADS in tons of methane per year, and we converted these to volumes of methane and then to natural gas, assuming a 78.8 percent methane content for

[1] Operators report royalties using a separate data collection form that operators are required to report to BOEMRE monthly.

natural gas.[2] In the GOADS study, fugitive emissions are estimated by looking at the number of valves and other components on a given production platform and then assuming an average leak rate. BOEMRE's data contractor performs a series of quality checks on the data after collection.

A third source of data on vented and flared volumes was a nationwide analysis performed by officials from EPA's Natural Gas STAR program, a national, voluntary program that encourages oil and gas companies, through outreach and education, to adopt cost-effective technologies and practices that improve operational efficiencies and reduce methane emissions. EPA's nationwide venting and flaring volumes were based on publicly available empirical data on national oil and gas production for 2006, 2007, and 2008, combined with knowledge of current industry practices, including usage rates and effectiveness of venting and flaring reduction technologies. For example, EPA used data on the number of well completions per year and data on the average venting per completion to estimate a yearly nationwide total from that source, with similar approaches used for estimating total venting and flaring from other key sources.[3]

EPA adjusted its estimates to account for the industry's efforts to control some venting and flaring emissions. EPA's analysis was limited in some ways, however. For instance, lacking empirical data on actual nationwide rates of use of certain control technologies, EPA based its analysis on anecdotal information in some cases. In order to be able to compare these data with the OGOR data, we scaled EPA's national estimates to federal leases based on the proportion of natural gas production on federal leases over total U.S. natural gas production using data from MRM and the Department of Energy's Energy Information Administration (EIA).[4] EPA also made estimates of offshore venting and flaring based on BOEMRE's 2005 GOADS data.

---

[2] We combined the pneumatic pump and pressure/level controller categories into our "pneumatic devices" category. We combined all the other source categories from GOADS into our "other sources" category.

[3] Due to incomplete data on oil storage tank emissions and reductions for 2008, the tank emissions from 2007 serves as an approximation for the emissions in 2008. EPA also included workovers with its estimates of venting and flaring from well completions. Workovers are remedial procedures designed to increase production on existing oil and gas wells.

[4] EPA's initial estimates of venting and flaring were for the methane component of natural gas. These volumes were converted to reflect overall natural gas emissions by assuming, for most sources, an average 78.8 percent methane content for the gas.

EPA officials adjusted volumes reported to GOADS based on publicly available information on current industry practices, including usage rates and effectiveness of venting and flaring reduction technologies.

A fourth source of venting and flaring data was based on analysis conducted by WRAP, a collaborative arrangement between tribal and state governments and various federal agencies set up to develop the technical and policy tools needed by western states and tribes to comply with the EPA's regional air quality regulations. As part of its efforts to better understand the oil and gas industry's impact on regional air quality, WRAP, through its contractor, the Environ International Corporation, collected data for 2006 on the volumes and sources of key air pollutants such as volatile organics and nitrogen oxides, which are associated with vented and flared gas. WRAP collected these data with backing from the Independent Petroleum Association of Mountain States, an industry group representing oil and gas producers in the western United States. We used Environ to reconfigure the data from the WRAP air quality analysis in order to estimate the overall volumes of vented and flared gas. The WRAP analysis focused on five specific production basins in the mountain west: the Piceance, Denver-Julesburg, and North San Juan Basins in Colorado; the Uinta Basin in eastern Utah; and the South San Juan Basin in northern New Mexico (see fig. 4). The WRAP analysis was based primarily on empirical data from operators in these basins, including drilling and production volume data, as well as data from a survey of operators. This survey asked operators to report actual vented and flared volumes, as well as to provide information on other aspects of their operations, including the emission control technologies they had in place. Similar to the EPA venting and flaring analysis, however, Environ did not have complete data from all operators in each basin and thus estimated some information based on survey data from a subset of operators.[5] In addition, the original WRAP data did not distinguish between federal and nonfederal oil and gas operations, so we provided federal well numbers to Environ so that they could identify the federal lease component of vented and flared gas.

To estimate the magnitude of potential increases in royalty payments and reductions in greenhouse gas emissions resulting from capturing additional vented and flared gas with available technologies, we had EPA provide us with estimates of the onshore expansion potential of a number of key technologies and associated venting and flaring volume reductions.

[5] BP provided most of the data for the North San Juan basin, and Environ was not able to verify its accuracy to the extent that it did for data reported in the other basins.

For simplicity, EPA developed these estimates by focusing on the expansion potential of a subset of technologies considered to provide the largest emission reductions. These estimates may be conservative, however, because they did not incorporate reductions from a number of other potential venting and flaring opportunities catalogued by the Natural Gas STAR program.[6] These estimates were not based entirely on comprehensive usage data collected from the oil and gas industry, but were based, in part, on publicly available evidence collected through years of experience with the oil and gas industry.

In addition, circumstances are constantly changing, and more technological innovations are potentially being used as time goes on, so there is some uncertainty in how much lost gas can be captured.

We also compared venting and flaring volumes and the types of emission-reduction technologies used in each of the basins from the WRAP data, allowing us to draw conclusions about the impact of different levels of technology on venting and flaring volumes.

We did not identify similar data on reduction opportunities offshore. We also interviewed officials from BLM, BOEMRE, EPA, and state agencies, as well as representatives from private industry, including technology vendors and an environmental consultant regarding the expanded use of available technologies to capture additional vented and flared gas. We conducted background research on venting and flaring reduction technologies, including from publicly available EPA Natural Gas STAR case studies. Finally, we obtained royalty information from MRM to calculate the royalty implications of the onshore venting and flaring reductions, and used conversion factors from EPA to calculate the greenhouse gas impacts of the vented and flared natural gas.[7] To assess the federal role in reducing vented and flared gas, we conducted interviews with officials from Interior, EPA, the Department of Energy, state agencies, and members of the oil and gas industry. We also reviewed agency guidance and documentation, other studies related to federal management and oversight of the oil and gas industry, as well as prior GAO

---

[6] It is difficult to estimate the extent to which the use of each control technology can be increased. Reductions may not always be feasible and depend on site-specific conditions. EPA's estimates of potential reductions from oil and condensate storage tanks also involved valve inspection and repair in addition to installing vapor recovery units.

[7] The conversion factor we used was .4045 million metric tons of carbon dioxide equivalent per billion cubic feet of vented natural gas, and 0.06 million metric tons of carbon dioxide per billion cubic feet of flared natural gas. We used a royalty rate of 11.45 percent and an average natural gas price of $4.01 per thousand cubic feet.

work that described limitations in the systems Interior has in place to track oil and gas production on federal leases.[8] We conducted interviews with officials in six BLM field offices (Farmington and Carlsbad in New Mexico; Vernal, Utah; Glenwood Springs, Colorado; Pinedale, Wyoming; and Bakersfield, California) and staff from BLM headquarters. We also interviewed BOEMRE staff in Denver, Colorado, and New Orleans, Louisiana.

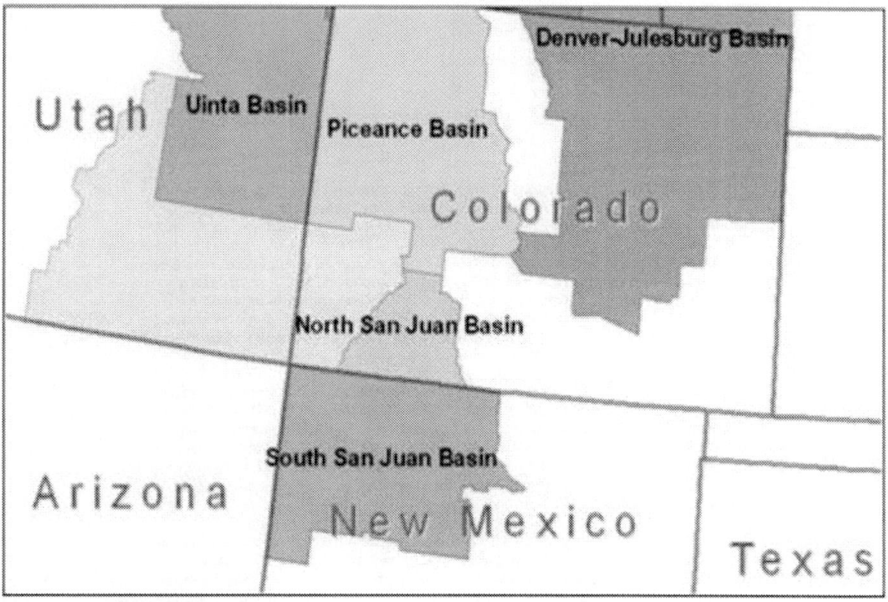

Source: WRAP/Environ.

Figure 4. Locations of Production Basins Included in WRAP Study.

We conducted this performance audit from July 2009 to October 2010 in accordance with generally accepted government auditing standards. Those standards require that we plan and perform the audit to obtain sufficient, appropriate evidence to provide a reasonable basis for our findings and conclusions based on our audit objectives. We believe that the evidence obtained provides a reasonable basis for our findings and conclusions based on our audit objectives.

---

[8] GAO, Oil and Gas Management: Interior's Oil and Gas Production Verification Efforts Do Not Provide Reasonable Assurance of Accurate Measurement of Production Volumes, GAO-10-313 (Washington D.C.: Mar. 15, 2010).

APPENDIX II: COMMENTS FROM THE DEPARTMENT OF THE INTERIOR

United States Department of the Interior

OFFICE OF THE SECRETARY
WASHINGTON, D.C. 20240

OCT 12 2010

TAKE PRIDE IN AMERICA

Mr. Frank Rusco
Director, Natural Resources and Environment
Government Accountability Office
441 G Street, N.W.
Washington, D.C. 20548

Dear Mr. Rusco:

Thank you for the opportunity to review and comment on the Government Accountability Office (GAO) draft report entitled, *Federal Oil and Gas Leases: Opportunities Exist to Capture Vented and Flared Gas, Which Would Increase Royalty Payments and Reduce Greenhouse Gases* (GAO-11-34). The draft GAO report includes five recommendations for the Secretary of the Interior that are intended to augment the present capability to capture vented and flared gas, potentially increase royalty payments, and reduce greenhouse gas emissions. More specifically, the GAO's recommendations address emissions data collection; the reconciliation of conflicting data; providing new emission reporting guidance to operators; and the need to explore using new technologies to identify and reduce volumes of flared and vented natural gas.

The Department of the Interior (DOI) concurs or partially concurs with all five recommendations. Within DOI, the Bureau of Land Management (BLM) and Bureau of Ocean Energy Management, Regulation and Enforcement (BOEMRE) have the major responsibilities for leasing Federal oil and gas and accounting for fluid mineral revenues. Responses to each recommendation are provided in the attached Enclosure. In addition, for your consideration, technical comments are being provided in a separate electronic transmission.

The DOI acknowledges GAO's concerns about the potential impacts from uncaptured vented and flared gas. As noted in your draft report, both BLM and BOEMRE took steps to minimize venting and flaring on Federal oil and gas leases following the 2004 GAO report on BLM and BOEMRE venting and flaring processes (GAO-04-809). Since the 2004 GAO report, both BLM and BOEMRE have issued guidance and regulations to limit venting and flaring of gas during routine operations. Current initiatives demonstrate the Department's continued efforts to improve its monitoring and reduction of vented and flared gas. For example, BLM is revising Onshore Orders, which will address use of new technology to reduce gas emissions. The BOEMRE published a final rule in the *Federal Register* on April 19, 2010, requiring operators on larger offshore platforms to route vented and flared gas from a variety of sources through a meter. This will enable operators to report flared gas separately from vented gas and allow better tracking of greenhouse gas emissions. Additionally, both BOEMRE and BLM will track the Environmental Protection Agency's greenhouse gas rulemaking and use venting and flaring reduction technology where it is economic.

As GAO reported in 2004, the Federal lands and waters already have among the best venting and flaring records in the world. The DOI is committed to managing venting and flaring from onshore and offshore Federal oil and gas leases to the extent of our authority and will continue to improve our programs to remain a world leader in oversight of venting and flaring.

We appreciate your suggestions for improving the regulatory oversight of Federal oil and gas. If you have any questions, please contact Andrea Nygren, BOEMRE Audit Liaison Officer, at 202-208-4343, or LaVanna Stevenson-Harris, BLM Audit Liaison Officer, at 202-912-7077.

Sincerely,

Wilma A. Lewis
Assistant Secretary
Land and Minerals Management

Enclosure

Enclosure

DOI Response to Government Accountability Office (GAO) Draft Report
Federal Oil and Gas Leases: Opportunities Exist to Capture Vented and Flared Gas, Which Would Increase Royalty Payments and Reduce Greenhouse Gases **(GAO-11-34)**

To help ensure that Interior's data is complete and accurate, the GAO recommends that the Secretary of the Interior direct the Bureau of Land Management (BLM) and the Bureau of Ocean Energy Management, Regulation and Enforcement (BOEMRE) to:

Recommendation 1: *Take additional steps to ensure that each agency has a complete and accurate picture of vented and flared gas, for both onshore and offshore leases, by (1) BLM developing more complete data on lost gas by taking into consideration additional large onshore sources and ways to estimate them not currently addressed in regulations – sources that EPA's newly proposed greenhouse gas reporting rule addresses and (2) BOEMRE reconciling differences in reported offshore venting and flaring volumes in OGOR and GOADS data systems and making adjustments to ensure accuracy of these systems.*

BLM: Concurs
BOEMRE: Concurs

The BLM will track the content and timing of the Environmental Protection Agency's (EPA) greenhouse gas rulemaking. The BLM will analyze options for new standards to track additional sources of natural gas released during drilling, testing, and production. The new standards will be incorporated in the new Onshore Order on waste prevention and beneficial use.

In the current BOEMRE environment, it is difficult to compare the volumes of gas vented and flared in Oil and Gas Operations Reports (OGOR) to the Gulfwide Offshore Activity Data System (GOADS), but reconciling the database differences may be feasible in the future. The functions and structure of the BOEMRE OGOR and the GOADS differ. The OGOR is a lease-based report that historically did not require vented and flared gas volumes to be separately reported. The GOADS is facility-based, and data collected and reported to GOADS separates venting and flaring volumes. In May 2010, the agency published new regulations requiring companies to install flare/vent meters on facilities that process more than 2,000 barrels of oil per day (BOPD), on average. The regulations also require all venting and flaring to be reported separately on the OGOR (even facilities processing less than 2,000 BOPD). As recommended in the 2004 GAO report (GAO-04-809), the Minerals Management Service (now BOEMRE) performed a cost-benefit analysis to determine the proper threshold at which companies should be required to use venting and flaring meters based on economic feasibility. The agency determined that it would be economically feasible for facilities that meet the 2,000 BOPD threshold. We expect that the new regulations requiring meters on these facilities will allow us to compare facility volumes reported on the OGOR and GOADS system, reconcile data on larger facilities, and improve the accuracy of data in both systems.

An important component for improving data accuracy will be educating the operators and reporters collecting and entering data into each system. The BOEMRE recently issued a Notice to Lessees (NTL) and Operators in the Gulf of Mexico announcing the GOADS 2011 effort. The NTL provided a GOADS users' guide and answers Frequently Asked Questions. It also announced a GOADS workshop, scheduled for October 2010, to discuss and explain the information collection and reporting procedures. In addition, the BOEMRE's Minerals Revenue Management program—now the Office of Natural Resources Revenue (ONRR)—issued a Dear Reporter letter in May 2010 to all operators, instructing them on how to report flaring and venting separately on the OGOR. The ONRR also conducts an average of four reporter training sessions a year. The ONRR will continue to educate operators on the flaring and venting reporting requirements, and in particular the new regulations, disposition codes, and meters needed to report correctly. This training will emphasize the need to accurately report flaring and venting on the OGOR and in the GOADS systems. The next reporter training will be held in February 2011. This will be a joint effort between the BOEMRE offshore and revenue management programs.

Once sufficient data are available and analyzed, BOEMRE can determine if additional reconciliation efforts are needed.

To help reduce venting and flaring of gas addressing limitations in its regulations [GAO] recommend[s] that the Secretary of the Interior direct BLM and BOEMRE to take the following four actions:

Recommendation 2: *BLM should revise its guidance to operators to make it clear that technologies should be used where they can economically capture sources of vented and flared gas, including gas from liquid unloading, well completions, pneumatic valves, and glycol dehydrators. BOEMRE should consider enforcing its requirement that gas be captured where economical for "lease-use" sources of gas.*

BLM: Concurs
BOEMRE: Partially concurs

The BLM will develop new standards to require use of new technologies that can economically capture vented and flared natural gas used in lease operations. The new standards will be incorporated in the new proposed Onshore Order on waste prevention and beneficial use.

With respect to BOEMRE, the GAO's recommendation 2 misapprehends the scope of the regulations governing "lease-use" sources of gas. Because BOEMRE has no present requirement to capture "lease-use" sources of gas where economic, new regulations would have to be implemented before such a requirement could be imposed and enforced. As mentioned in your report, GAO was unable to quantify the volume of natural gas that would be saved by such a regulatory change. It is highly uncertain at this time if the volume of gas saved would be large enough to make a capture requirement economic. Thus, BOEMRE will evaluate the issue further and determine if new regulations are warranted.

Recommendation 3: *BLM and BOEMRE should assess the potential use of venting and flaring reduction technologies to minimize the waste of natural gas in advance of production where applicable, and not solely for purposes of air quality.*

BLM: Concurs
BOEMRE: Concurs

The BLM will analyze options for new standards to require use of venting and flaring reduction technologies, where applicable. The new standards will be incorporated in the new Onshore Order on waste prevention and beneficial use.

The BOEMRE diligently addresses air quality in advance of production. The meter requirement stemming from new regulations will provide data on the extent of venting and flaring. Once sufficient data exist, BOEMRE will use the data collected from the new meters to analyze the need and feasibility of requiring venting and flaring reduction technologies on facilities in advance of production, for purposes other than air quality.

Recommendation 4: *BLM and BOEMRE should consider the expanded use of infrared cameras where economic to improve reporting of emission sources and to identify opportunities to minimize lost gas.*

BLM: Concurs
BOEMRE: Concurs

The BLM agrees that infrared cameras can be a useful tool for detecting natural gas emissions as part of a directed inspection program. The BLM will implement the use of infrared cameras during inspections to spot check emission sources and minimize lost gas.

The BOEMRE will expand its use of infrared cameras where economic.

Recommendation 5: *BLM should collect information on the extent that larger operators use venting and flaring reduction technology and periodically review this information to identify potential opportunities for oil and gas operators to reduce their emissions, and BOEMRE should use existing information in its GOADS data system for this same purpose, to the extent possible.*

BLM: Concurs
BOEMRE: Concurs

The BLM will analyze options to collect information on how larger operators use venting and flaring reduction technology on Federal onshore leases, and periodically review this information to identify potential opportunities for oil and gas operators to reduce their emissions. The options adopted will be incorporated in the new Onshore Oil and Gas Order on waste prevention and beneficial use.

The BOEMRE will use the meters required by the new regulations to gather data on offshore venting and flaring volume, and periodically review and identify potential opportunities for oil and gas operators to reduce their emissions. The BOEMRE will also evaluate the 2008 data in GOADS to identify potential opportunities where reduction technology could be economically applied under the Outer Continental Shelf Lands Act conservation mandate. The BOEMRE will track EPA's greenhouse gas rulemaking and use venting and flaring reduction technology where it is economic.

APPENDIX III: VOLUMES AND SOURCES OF VENTED AND FLARED GAS BASED ON ANALYSIS OF 2006 WRAP DATA

Table 5. Volumes and Sources of Vented and Flared Gas from the Uinta Basin

Sources from Uinta Basin	Volume (Bcf)
Pneumatic devices	4.3
Glycol dehydrators	4.3
Fugitive emissions	0.4
Oil and condensate storage tanks	0.2
Other sources	0.1
Total	9.2

Source: Environ Corp. analysis of 2006 WRAP data.
Note: Volume figures in table may not sum to totals due to rounding.

Table 6. Volumes and Sources of Vented and Flared Gas from the North San Juan Basin

Sources from North San Juan Basin	Volume (Bcf)
Glycol dehydrators	0.053
Gas well liquid unloading	0.004
Flaring	0.003
Pneumatic devices	0.001
Fugitive emissions	0.001
Total	0.062

Source: Environ Corp. analysis of WRAP data.
Note: Volume figures in table may not sum to totals due to rounding.

Table 7. Volumes and Sources of Vented and Flared Gas from the South San Juan Basin

Sources from South San Juan Basin	Volume (Bcf)
Well completions	5.3
Glycol dehydrators	3.2
Gas well liquid unloading	1.9
Fugitive emissions	0.6
Pneumatic devices	0.4
Other sources	0.1
Total	11.6

Source: Environ Corp. analysis of WRAP data.
Note: Volume figures in table may not sum to totals due to rounding.

Table 8. Volumes and Sources of Vented and Flared Gas from the Denver-Julesburg Basin

Sources from Denver-Julesburg Basin	Volume (Bcf)
Oil and condensate storage tanks	0.035
Pneumatic devices	0.030
Fugitive emissions	0.019
Gas well liquid unloading	0.006
Well completions	0.002
Other sources	0.004
Total	0.095

Source: Environ Corp. analysis of WRAP data.

Note: Volume figures in table may not sum to totals due to rounding.

End Notes

1 In June 2010, the Secretary of the Interior changed the name of the Minerals Management Service to BOEMRE.

2 For the purposes of this report, we use the term "natural gas" to mean the mixture of gas resulting from oil and gas production activities. This natural gas will vary in content but, on average, is approximately 80 percent methane, with the remaining 20 percent a mix of other hydrocarbons and nonhydrocarbons, such as carbon dioxide and nitrogen.

3 Major greenhouse gases include carbon dioxide (CO_2); methane (CH_4); nitrous oxide (N_2O); and synthetic gases such as hydrofluorocarbons (HFC), perfluorocarbons (PFC), and sulfur hexafluoride (SF_6).

4 GAO, *Natural Gas Flaring and Venting: Opportunities to Improve Data and Reduce Emissions*, GAO-04-809 (Washington D.C.: July 14, 2004).

5 Much of our information and data about offshore leases from BOEMRE came from its Offshore Energy and Minerals Management program.

6 EPA developed these estimates to support a proposed rule in 2010 to require the reporting of greenhouse gas emissions from the oil and gas industry. These estimates were for 2006 to 2008.

7 WRAP is a collaborative effort of tribal governments, state governments, and various federal agencies to address western air quality concerns. It is administered by the Western Governors' Association and the National Tribal Environmental Council.

8 Independent Petroleum Association of Mountain States is an industry association representing oil and gas producers in the western United States.

9 Production basins are land formations with subsurface oil and natural gas reservoirs, often covering hundreds of square miles.

10 GAO, *Oil and Gas Management: Interior's Oil and Gas Production Verification Efforts Do Not Provide Reasonable Assurance of Accurate Measurement of Production Volumes*, GAO-10-313 (Washington D.C.: Mar. 15, 2010).

11 Pub. L. No. 66-146, 41 Stat. 437 (1920).

12 67 Stat. 462 (1953) codified at 43 U.S.C. § 1331 *et seq.*

[13] These routine amounts are laid out in BLM and BOEMRE regulations and guidance and allow certain volumes from a number of operations, such as well completions. Operators are required to notify these agencies if they plan to intentionally vent and flare beyond routine amounts. BLM and BOEMRE then classify the gas as either unavoidably or avoidably lost based on their judgment of the technical or economic feasibility of capturing the gas, and royalties are due on losses deemed avoidably lost.

[14] While not considered "vented" gas by Interior, we include fugitive emissions in this report as a relevant source of lost gas.

[15] Condensates are hydrocarbon liquids that are byproducts of natural gas production. Unprocessed natural gas normally contains a small amount of water vapor, and dehydration equipment is used to remove this moisture prior to pipeline transportation.

[16] Older detection technology consisted of using manual probes, which need to be very close to the venting source.

[17] This video shows vented gas, which appears to be smoke, billowing from the top of cylindrical metal oil storage tanks and from a pneumatic valve. Video clips courtesy of EPA and a private emission detection firm.

[18] The meters used by BOEMRE and BLM to establish production volumes, upon which royalty payments are based, are generally located downstream of the production site, at the point where the oil and gas enter the sales pipeline or other distribution network.

[19] The Intergovernmental Panel on Climate Change is the body for the assessment of climate change, established by the United Nations Environment Programme and the World Meteorological Organization to provide a scientific view on the current state of climate change and its potential environmental and socio-economic consequences.

[20] EPA defines volatile organic compounds as certain carbon compounds that participate in atmospheric chemical reactions.

[21] EPA is responsible for the regulation of such air pollutants and has been recently charged with regulating greenhouse gas emissions under the Clean Air Act.

[22] BOEMRE has primary regulatory responsibility over air emissions from offshore sources in the central and western Gulf of Mexico.

[23] See EPA's Natural Gas STAR Web site for more information on these technologies (www.epa.gov/gasstar/).

[24] Triethylene glycol is the active chemical in the operation of this equipment. A flash tank separator is a device that captures additional gas from glycol dehydrators.

[25] GAO-04-809.

[26] BLM determined that requiring thousands of onshore operators to install meters would be prohibitively expensive.

[27] The Department of Energy's EIA is responsible for producing independent, unbiased research that helps the public, the federal government, industry, and the Congress better understand energy markets and promote sound policy making. EIA collects and analyzes data on the supply, consumption, and prices of oil and gas.

[28] EPA's estimates were based on publicly available oil and gas production data and information collected from industry participants in the Natural Gas STAR program, a nationwide, voluntary effort spearheaded by EPA aimed at reducing methane emissions from the oil and gas industry.

[29] Operators are required to report the sum of their vented gas and flared gas; they are not required to identify individual sources of lost gas.

[30] BLM's guidance was included in Notice to Lessees and Operators (NTL) 4A. Interior issues NTLs to clarify existing regulations. Operators need permission to vent or flare above

routine amounts. In dealing with vented and flared gas, BLM's key guidance is in the form of an NTL. Offshore, BOEMRE uses regulations to guide operators in addressing vented and flared gas.

[31] Additional guidance from MRM explains how operators should submit data to the OGOR system, but does not provide detail on which sources to report, or on how they should be estimated.

[32] These limits generally allow operators to vent and flare gas required for routine well operations.

[33] See appendix I for more detail on how EPA developed its estimates.

[34] It is possible to estimate venting and flaring based on known emission rates of equipment type or production method. For example, if a pneumatic device is known to vent 10 cubic feet per hour, an operator would multiply that rate by the number of hours the piece of equipment operates each day.

[35] See appendix III for more details on the volumes and sources from the other four basins.

[36] See appendix I for more detail on the development of the estimates based on WRAP data.

[37] Offshore gas production in the Gulf of Mexico made up about 98 percent of total federal offshore natural gas production in 2008.

[38] Lease-use, or beneficial use, gas refers to natural gas that BLM and BOEMRE allow operators to use to power oil and gas production equipment on the lease. Emissions from pneumatic devices and glycol dehydrators would have been reported as lease-use gas, according to BOEMRE officials, and we determined it was not possible to extract these volumes from OGOR because they were combined with a number of other nonvented sources.

[39] The GOADS study included estimates of losses from natural gas compressors, although EPA's estimates were greater because of higher assumed losses from the compressor seals.

[40] BOEMRE officials told us that they did not draw conclusions from comparisons between 2005 GOADS and OGOR data because of the effect Hurricane Katrina and Rita had on offshore production in that year.

[41] Texas Commission for Environmental Quality (TCEQ), *Upstream Oil and Gas Storage Tank Project Flash Emissions Model Evaluation* (July 16, 2009).

[42] GAO-04-809.

[43] 75 Fed. Reg. 20291-20293 (April 19, 2010). The rule also requires vented gas and flared gas to be reported separately.

[44] For simplicity, EPA developed this estimate by focusing on the expansion potential of a subset of technologies considered to provide the largest emission reductions. The estimates may be conservative, however, because they did not incorporate reductions from a number of other potential venting and flaring opportunities catalogued by EPA's Natural Gas STAR program.

[45] Although there is likely some chance for similar reductions offshore, EPA did not estimate this amount.

[46] For more detail on carbon offsets see GAO, *Carbon Offsets: The U.S. Voluntary Market Is Growing, but Quality Assurance Poses Challenges for Market Participants*, GAO-08-1048 (Washington, D.C.: Aug. 29, 2008).

[47] For these calculations, we assumed average onshore royalty payments of 11.45 percent, the average onshore royalty rate in 2009. See appendix I for more details.

[48] Methane breaks down in the atmosphere more quickly than CO_2 and lasts an average of 12 years in the atmosphere. This accounts for its greater impact over the shorter time frame.

[49] This statement assumes that venting and flaring are reduced in proportional volumes.

[50] Overall, according to EPA's analysis, in addition to the total potential federal reductions of 50 Bcf, nonfederal wells could have added an additional 252 Bcf in reductions with more widespread use of venting and flaring reduction technologies.

[51] For onshore leases, *see* 43 C.F.R. § 3161.2. For offshore leases, *see* 30 C.F.R. § 250.106.

[52] BLM guidance is in the form of a Notice to Lessees and Operators (NTL). According to the NTL, the operator can vent or flare gas during operations such as clearing the drilling waste or removing liquid from the well for 24 hours without obtaining permission from BLM to vent gas. The operator may also flare or vent any gas vapors released from storage tanks or low pressure production vessels unless BLM determines that the recovery of the gas would be warranted. The vented or flared gas is considered to be "unavoidably lost."

[53] If an operator does not exceed these limits—which is almost always the case according to BLM staff—BLM does not consider the economic and technical viability of further reducing venting and flaring. BLM inspectors also note obvious signs of vented and flared gas during their inspections, which occur at least every 3 years, and try to verify that operators have permission for the release.

[54] In a previous report, we found that BOEMRE had not met these annual inspection goals. See GAO, *Oil and Gas Management: Interior's Oil and Gas Production Verification Efforts Do Not Provide Reasonable Assurance of Accurate Measurement of Production Volumes*, GAO-10-313 (Washington D.C.: Mar. 15, 2010).

[55] 43 C.F.R. § 3161.2.

[56] So far there has been only one rural oil and gas production basin, the Jonah-Pinedale basin in Wyoming, that is not meeting EPA standards for ground-level ozone.

[57] Applications include detailed information on plans for drilling and completing wells, such as the amounts and types of cement used, the construction materials, the methods for handling waste, the plans for surface reclamation, and multiple other subjects for BLM to consider. In addition, the operator submits a diagram of existing or proposed production facilities.

[58] 30 C.F.R. § 250.246.

[59] According to one BLM official we spoke with, an inspector may note whether or not operators use particular types of venting and flaring equipment, but the field office does not keep specific records about equipment use.

[60] One consulting firm we spoke with used infrared cameras to detect leaks on clients' offshore oil platforms and found, on average, 21 gas leaks per facility, totaling an estimated 127,000 cubic feet of gas per day.

[61] In addition to the Natural Gas STAR program, EPA's Office of Research and Development is developing specialized measurement approaches to remotely detect and quantify air emissions, including methane, from the oil and gas industry and other sources.

In: Flaring and Venting of Natural Gas
Editor: Linda R. Henderson
ISBN: 978-1-63463-838-8

Chapter 3

TESTIMONY OF ANDREW LOGAN, DIRECTOR, OIL & GAS INDUSTRY PROGRAM, CERES. HEARING ON "BENEFITS OF AND CHALLENGES TO ENERGY ACCESS IN THE 21ST CENTURY: FUEL SUPPLY AND INFRASTRUCTURE"*

Chairman Whitfield, Ranking Member Rush, members of the subcommittee, thank you for the opportunity to be here today to testify on the economic and environmental impacts of natural gas flaring in the United States.

Ceres is a non-partisan, non-profit organization. We are a coalition of institutional investors and environmental organizations working to make capital markets more environmentally and socially sustainable. We have over 100 institutional investor members representing over $11 trillion in assets, united by the belief that strong environmental and social performance drives strong financial performance over time.

Our investor members have significant financial exposure to the oil and gas sector, and very much want to see the industry succeed. While shale oil is bringing significant economic benefits to the United States, we believe that the way the resource is currently being developed is extremely short-sighted, and is failing to capture its full value.

---

* This is an edited, reformatted and augmented version of testimony presented March 6, 2014 before the House Energy and Commerce Committee, Subcommittee on Energy and Power.

Our investors believe that flaring natural gas is environmentally destructive, economically wasteful and, most importantly, it is almost always unnecessary. And despite well-intentioned and quite significant efforts by some companies, the problem is getting worse, and will continue to get worse until the regulatory environment changes so that flaring is no longer the cheapest and easiest option.

Flaring is a problem that the US thought it had left behind in the 1950s. But the rapid growth of tight oil production in the United States, in places like North Dakota and Texas, has been accompanied by a dramatic increase in flaring that has propelled the U.S. into the top 10 gas flaring countries globally along with Russia, Nigeria, and Iraq.[1] This is not the sort of company that the US should be keeping.

Most of this flaring is occurring at oil wells drilled in areas that lack the pipeline and processing infrastructure necessary to capture the gas that comes out of the ground with the oil. Instead of investing in the infrastructure necessary to capture that gas, companies are often choosing to simply flare it off where current regulations allows them to do so. It is important to note, though, that lack of infrastructure is only part of the problem. Roughly half of all flaring in the Bakken comes from wells that are already connected to pipelines.[2] So we need better planning as well—we want to see industry plan its wells with the idea that natural gas has value.

This is coming at a steep environmental cost. Flaring is a major contributor to greenhouse gas emissions. In 2012, the emissions from the flared gas in North Dakota alone were equivalent to adding over one million cars to the road.[3] In addition, because the flares used often only partially combust the natural gas, a variety of other hazardous pollutants are generated by the process, including black carbon, another potent driver of climate change with adverse health effects.

While flaring in North Dakota has received the most attention, it is a problem nearly everywhere that has seen significant tight oil production. Texas has seen a 10-fold increase in flaring permits since 2010, and flaring is a growing concern in Wyoming as well.

The environmental impact of flaring is not its sole cost. North Dakota gas is so rich in valuable natural gas liquids like propane and butane that this is about the last gas you would want to flare. In fact, over the course of 2012, North Dakota oil and gas producers flared more than $1 billion of natural gas, a massive economic waste.[4]

So flaring is clearly environmentally damaging. It is economically wasteful, to the tune of a billion dollars a year- and growing- in North Dakota alone. But most importantly, it is avoidable.

The North Dakota Industrial Commission has run the numbers, and has concluded that it *is* economic to capture this gas, in large part due to its high liquids content.[5] But flaring in the state is still at around a third of the total gas production. And that is because, while capturing gas produces positive economic returns, it doesn't match the returns from drilling the next oil well. So if regulations allow that sort of short-term decision making, as they do in North Dakota, many companies will make that choice.

Our investors take a long-term view and want to see the value of the resource maximized. They are deeply concerned by the current approach to development. The Bakken formation has been around for 360 million years. It's not going anywhere. If it takes a little extra time to develop the resource in a thoughtful and deliberate way, it seems to me we should strongly encourage that.

We are working with our investors to push the industry to take a longer-term view, and it is important to acknowledge that some companies are doing so. Continental, the biggest producer in the Bakken, is sharing drilling plans with pipeline companies several years in advance to allow infrastructure to get a head start, and has managed to bring its flaring rate down to 10%. Hess is planning its drilling program to ensure wells can be quickly connected to gas gathering lines and has made investments in gas processing infrastructure. Yet the data are clear: the problem is getting worse, not better. Flaring in North Dakota hit 36% in December, a new record. This means that more than 1/3 of all natural gas produced in the state is going up in smoke, at the same time as consumers around the country are seeing price spikes from natural gas in this cold winter, along with actual shortages of propane in many places.

Flaring is an indefensible economic waste, but it also represents a major opportunity- a $1 billion/year opportunity for entrepreneurs as well as the industry itself. We are seeing huge amounts of innovation going on, with companies from start-ups to General Electric developing technologies to utilize gas at the wellhead or get it to market without pipelines. There is the potential for a real American success story here. But this technology is having a hard time gaining a foothold because it's hard to compete with free. And right now in North Dakota, flaring is free. If you take only one point away from my testimony today, it's that it shouldn't be.

SUMMARY OF TESTIMONY OF ANDREW LOGAN DIRECTOR, OIL & GAS INDUSTRY PROGRAM CERES

1. Flaring of associated natural gas is a growing problem in the United States
   a) The flaring rate in North Dakota hit a record high of 36% in December, the most recent month for which data is available
   b) Texas has seen a 10-fold increase in flaring permits since 2010, and flaring is a growing concern in Wyoming as well.
   c) Despite significant investment, and leadership by a handful of companies, the problem is only getting worse and will continue to get worse until the regulatory environment changes so that flaring is no longer the cheapest and easiest option.
2. Flaring is environmentally destructive...
   a) In 2012, the emissions from the flared gas in North Dakota alone were equivalent to adding over one million cars to the road.
   b) In addition, because the flares used often only partially combust the natural gas, a variety of other hazardous pollutants are generated by the process, including black carbon, another potent driver of climate change with adverse health effects.
3. Economically wasteful...
   a) a. In 2012, North Dakota oil and gas producers flared more than $1 billion of natural gas, a massive economic waste.
4. And largely avoidable
   a) According to the North Dakota Industrial Commission, it is economic to capture Bakken gas, in large part due to its high liquids content. But flaring in the state is still at around a third of the total gas production. And that is because, while capturing gas produces positive economic returns, it doesn't match the returns from drilling the next oil well. So if regulations allow that sort of short-term decision making, as they do in North Dakota, many companies will make that choice.
5. Flaring will only be solved when the regulatory structure changes so that flaring is no longer the easiest option. This holds the potential to unleash massive innovation, and capture a $1 billion/year market opportunity
6. At the moment, flaring regulation is done mostly at the state level. There is an option for federal oversight by expanding EPA New

Source Performance Standards, which currently only cover natural gas wells, to include oil wells.

End Notes

[1] Global Gas Flaring Reduction "Estimated Flared Volumes from Satellite Data, 2007-2011" http://go.worldbank.org/D03ET1BVD0

[2] Fielden, Sandy "Set Fire To The Gas– The Fight to Limit Bakken Flaring" 6 May 2013, https://rbnenergy.com/set-fire-to-the-gas-the-fight-to-limit-bakken-flaring

[3] Salmon, Ryan "Flaring up: North Dakota Natural Gas Flaring More Than Doubles in Two Years" http://www.ceres.org/resources/reports/flaring-up-north-dakota-natural-gasflaring-more-than-doubles-in-two-years/view

[4] ibid

[5] Helms, Lynn "Director's Cut" https://www.dmr.nd.gov/

INDEX

D

E

F

L

M

N

O

P

V

W

Y